Time & Clocks
for the Space Age

James Jespersen
& Jane Fitz-Randolph

Time
&
Clocks
for the Space Age

ILLUSTRATED WITH

PHOTOGRAPHS & DIAGRAMS

Atheneum   1979   New York

*The authors would like to thank Dr. James A. Barnes and
Collier M. Smith of the National Bureau of Standards for
reading the manuscript and making many helpful suggestions.*

LIBRARY OF CONGRESS CATALOGING IN PUBLICATION DATA

Jespersen, James.
Time and clocks for the space age.

Includes index.

SUMMARY: Discusses the history of time measurement
and the evolution of measuring efforts and devices,
describes ways of using precise time,
and outlines some of the scientific studies in which
time plays a significant role.
1. Time—Juvenile literature.   2. Clocks and
watches—Juvenile literature.   [1. Time.
2. Clocks and watches]   I. Fitz-Randolph, Jane,
joint author.   II. Title.
QB209.5.J47   529'.7   79-14578
ISBN 0-689-30710-1

*To
Ann,
Christine,
and Andrew;
Fred,
Philip,
and Priscilla*

Contents

III Time & Science

I

From
Sundials
to
Atomic
Clocks

1
What Is Time?

WE ALL KNOW what we mean when we talk about time. Even small children know when it's playtime, bedtime, dinnertime, time to look at a TV program. We all try to save time and not to lose it. We keep time with the band and make time for things we want to do. We waste time, beat time, and sometimes kill time. Yet there's always more of it—the same amount for everyone.

But what *is* time? This seemingly simple question has puzzled the world's great thinkers for centuries, from Plato and Galileo to DesCartes, Newton, and Einstein. Sir Isaac Newton could say only that "time is that which passes," and Dr. Albert Einstein concluded that "time is what a clock measures." Today's scientists are still unable to agree on a definition that suits them. Engineer/philosopher Buckminster Fuller, famous for his geodesic dome, says that "time is something we wait in."

Whether they can define it or not, scientists agree that time is a physical quantity which, like other physical quantities such as mass and temperature, is a necessary part of many scientific and mathematical equations. To find out when we may expect to arrive at a nearby city, for example, we need to know how many *kilometers per hour* we will

3

average on our journey. And the instructions on a box of cake mix tell us *how many minutes* to bake the cake at a specific *temperature*.

Yet time is different from other physical quantities in some strange ways. We can see length and we can feel weight and temperature. But none of the physical senses can tell us anything about time. We cannot see, hear, feel, smell, or taste time. We know it only as an idea, or by observing its effects. As we watch the sun nearing the horizon, for example, we know it it getting "later" in the day.

Another peculiar thing about time is that it "passes," and it moves in only one direction. We can measure the length of a football field beginning at either end and moving "forward" to the other end. We can measure the amount of molasses in a barrel by dipping the molasses from the top or by drawing it from a spigot at the bottom. But we must always think of time as now, before now, and after now. And "now" is constantly changing. Furthermore, the only time we can do anything is "now." No one can do anything yesterday or next week—until next week becomes "now."

Perhaps it's little wonder that some scientists aren't sure that time even exists. But whether it does or not, we shall undoubtedly go on "keeping" it, building devices for measuring it, and acting as if it does.

One thing that makes it hard to define time is that the one word *time* has two distinct meanings: Sometimes we want to know the *time when* something happened, or the "date"; and other times we want to know *how long a time* something lasted, or the "time interval" it covered. Kickoff time for the football game may be 2:00 P.M. on October 19. The *date* is important if we want to see the game. *How long* the game lasts doesn't matter much to the spectators, but the timekeeper responsible for timing the plays and the quarters has a very great interest in *time interval*.

It may seem that these two ideas should go together like a pair of shoes. But this is not the case, as we shall see. It's more like trying to make a pair out of one black and one brown shoe. The two ideas are often at odds; and the better the clocks we build and the more our communication and transportation industries grow, the more important the distinction becomes.

Both date and time interval can be measured very precisely, to a small fraction of a second. An event may begin September 4, 1980, 15h, 23m, 41.24s; *h, m,* and *s* mean hours, minutes, and seconds; the 15th hour, on a 24-hour clock, would mean 12 plus 3, or three o'clock in the afternoon. The event may last for a billionth of a second or less, or for a week or a year or a century or more.

To add to the problems of satisfying the needs of both those interested in date and those interested in time interval, both often want to be able to *synchronize* their measurements very precisely with someone else's measurements. Synchronize is a Latin word that means literally "time together." Two pigeon racers, for example, separated by many miles, may wish to release their birds at exactly the same instant. This means that two or more *clocks* must tell exactly the same *date* at the same moment. Sometimes clocks must agree not just within a second or two, but within *millionths* of a second. There's a "clock," for instance, in a TV broadcasting station that "sees" what the camera sees. And it has to tell the "clock" in your home TV set "what time it is" 15,000 times every second! If these *synchronization pulses* did not keep the two clocks running precisely in step, there could be no picture.

But we're getting ahead of our story. Let's go back and take a closer look at this strange, three-headed beast we call time. If you and your friend in another state are planning a camping trip and you want to know how long his canoe is, you can write him a post card and ask him. And the answer you receive on another post card tells you what you want to know. Or you might ask him how much his bowling ball weighs or how large his swimming pool is.

But what if you ask him for the time? He listens carefully for the radio announcement, perhaps, and writes the time on a post card for you. But even before he starts to write, what he would put down is no longer true and the information is useless.

Measuring time presents almost as many thorny problems as trying to define it, because it won't stand still. A chemist has a balance and a set of weights for measuring precisely the weight or mass of various things he uses. When he is not using the weights, he can put them away in a box and forget them until he needs them again. Similarly, a draftsman has a set of precise measuring instruments that he takes out whenever he needs them, knowing they are always the same, always ready for use.

But the scientist who needs to measure time has no such convenient tools. True, he may have a very fine clock or watch—the instruments for measuring time. But his measuring stick takes constant care. He has to keep it "running," for if it "stops"—or if it runs inaccurately—he "loses" the time. If he "starts" his clock again, he may be able to measure *time interval*, but he can regain the *date* or "time of day" only by referring to some other time source and then "resetting" his own.

Another difficulty is that there's really no way for him to tell, simply by looking at a clock or watch, whether it's "right" or not. This is a constant problem even for watch salesmen and repairers, especially now that there are so many electronic watches that keep time quite precisely. They would like to hand the customer a watch that is correct to the fraction of a second, but they can never be sure that even the best clock they have in their shop is correct. It may have lost a fraction of a second—or much more—since they last checked an hour ago or even a few minutes ago. Many jewelers and watch repairers do, in fact, contact the National Bureau of Standards time service daily or several times a day, to check their own clock that they use as a standard and the watches they repair and sell.

But again we're getting ahead of our story. We can depend on a meter stick or gram weight, once we know its accuracy, to keep on being accurate. But a clock or watch may "run" fast or slow by a little or a lot, and there's no way to know until we compare it with another clock. Even then we can't be sure whether it's our own clock or the other one that's wrong—unless we've checked with a clock much better than our own. We may have to turn to a third one. But then what if it doesn't agree with either of the other two?

How *do* we measure time? How can we know which clock is exactly right? And how can scientists keep clocks all over the world and on ships at sea and satellites in outer space telling the same time every second of every day and night? The answers to these and other questions give us a fascinating story of mankind's efforts, from very ancient times to the present, to understand this peculiar physical quantity called time, to keep track of it and to get a handle on it so that they can use it—in hundreds of ways that most of us never think about.

2

What Is a Clock?

SINCE WE'VE SEEN what a peculiar thing time is, we should not be surprised to find that one of its characteristics that makes it hard to measure also makes it possible to measure, very precisely. Thousands of times more precisely than we can now measure length or mass.

We've just noted that time "passes," and that we must always think of time as now, before now, and after now. So suppose we think of "now" as a kind of gate, with time passing through it at a steady rate like an endless string of beads. If we could stand at this gate and count the beads as they pass through, and keep track of our count, then we could measure time very nicely.

This is, in fact, the way we do measure time. The "beads" may be large—such as months or years—or they may be so small that it takes billions of them to make one second. But large or small, they pass through the "now" gate just one at a time, and we can count them. In its broadest meaning, that's what a clock is—a device for counting beads or units of time.

What are these beads, and where can we find them? Well, there are several possibilities. What we need is something that happens over and over again in the same way—some "cycle" that exists in nature perhaps. Certainly one of the first things we think of is the sun as it rises each morning and travels across the sky. We might also think of phases of the moon, the changes in the positions of stars and planets, and the cycles of seasons.

These things provided primitive peoples with the only clocks they had. Their "scientists" must have spent many thousands of hours, over long periods of many years, studying the movements of the sun, moon, planets, and stars. They kept records of their observations, and finally they were able to predict very accurately when eclipses and other astronomical events would happen. These astronomical clocks—especially the earth-sun clock—are still the basis for our measurements of time on earth.

But months and years—even days—are rather large beads of time. Anyone who wishes to practice playing the piano for an hour or watch a certain TV program at 5:30 P.M. will have to find some way to break the day bead into smaller beads. He might do this by standing a stick in the sand, and then watching and measuring when its shadow was shortest. He could mark a line here for "noon." Then by dividing the spaces on both sides of noon according to rather complicated astronomical principles, he could keep track of hours, and even minutes and seconds. In other words, he could make a sundial. The word *dial* comes from the Latin word for day—dies.

Although most sundials that we see today in parks and gardens are mainly ornamental, carefully designed and accurately placed sundials are very good clocks—when the sun is shining. A sundial "loses" the time every night, but displays it accurately and without attention from anyone again the next day. Such a clock never runs down or wears out. Some very old sundials, made centuries ago, tell time just as well today as they did when they were new. There were even sun "watches" that one could carry in his pocket, and some of these are still made today.

A sundial is a good clock when the sun is shining.

But the finest sundial is useless at night or on a cloudy day. And even on bright days, it's often a nuisance to have to go out into the sunshine just to find out what time it is. There are other disadvantages, too. A sundial tells the time only where it is; two sundials even 10 meters apart east and west tell slightly different time. (All sundials on a line running north and south would, of course, show noon at the same time. Do you see why?) Synchronizing sundials with each other is not practical.

For these and other reasons people throughout the centuries and all over the world have given a great deal of thought and effort to finding ways to measure time without having to refer to the sun itself. Until very recently the clocks they devised were mostly mechanical instruments of some kind. They used such things as sand or water flowing from one vessel to another or the step-by-step "ticking" movement of cog wheels driven by a weight or spring as their basis. We shall discuss some of these early clocks in the next chapter. A big breakthrough came in 1656 with the first pendulum clock, built by a Dutch scientist named Christian Huygens.

But all of these clocks had serious limitations that no

purely mechanical device could overcome. So scientists looked for some completely new way to make a clock. They knew they had to have three things that are essential to all clocks:

1. There must be some device that repeats some kind of activity over and over in "cycles" for a fairly long period of time. This device is called a *resonator*. A swinging pendulum is an example.

2. There must be some source of energy to keep the resonator going, such as a weight or a spring. The resonator and the energy source together produce an endless string of beads of exactly the same size and evenly spaced so that they will pass through the *now* gate at exactly the same rate or *frequency*. The resonator with its energy source is called a *frequency standard.*

3. There must be some way to count, keep track of, and display the number of cycles or beads of time made by the frequency standard as they pass through the *now* gate. The face and hands of a common wall clock, for example.

The clocks and watches we know best are made to display time for 12 hours and then start over. Often there's a second hand that counts seconds up to one minute. There are 24-hour clocks in some laboratories, and some clocks and watches also tell the day of the month. Stop watches tell only time *interval* in tenths of a second, up to 30 minutes or an hour perhaps.

But what if one wishes or needs to measure time interval in much smaller beads than this? In hundredths, thousandths, millionths, or even billionths of a second? The shutter on many cameras can be set to operate at 1/200 or 1/500 second. How can the manufacturer be sure this operation is accurate? Power companies in the United States supply electricity to our homes at a very reliable 60 cycles per second—or 60 hertz (Hz) , as it is called today. If they missed by even 5 hertz, our common electric clocks would not keep time; and there would be many other problems, including massive blackouts that might cover several states. Radio and television stations must have frequency standards that are accurate to one part in a million to keep their broadcast at the proper frequency. These are only a few of the commonest needs for clocks or frequency standards that can provide very small beads of time.

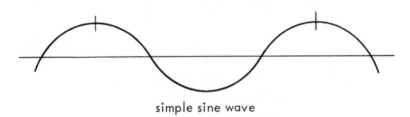

simple sine wave

Clocks and frequency standards are not the same thing, although both measure time, and frequency standards are often loosely called clocks; a clock's works, without its hands and face, is a frequency standard. It's easy to see that if we have a frequency standard and a way to count the beads as they pass through the *now* gate and display the count, then we also have a clock that can tell us both time interval and the date very precisely. Of course, we can keep track of the date only if we have some way to know when to start counting. But this is also true of the finest watch or clock.

But what kind of resonator vibrates hundreds or thousands or millions of times a second in a way that we can depend on? One answer came in 1929, when an American scientist, Dr. Warren A. Marrison, built the first clock that used a thin crystal of quartz rock as a resonator. He made it vibrate by applying an electric current to it. The first quartz-crystal clocks were enclosed in cabinets about 3 meters high, 2½ meters wide, and 1 meter deep, to hold all the necessary parts. Today, miniaturization of electronic circuits has made it possible to put all these things into a *wrist watch*. The tiny quartz crystal, about the size of a match head, is powered by a battery about the size of a dime.

How many beads pass through a quartz-crystal *now* gate in a second depends on the thinness to which the crystal is ground. Typical frequencies are 2.5 or 5 megahertz (MHz)—2½ or 5 million beads per second. The beads are counted very precisely, electronically, of course.

But even these beads are not small enough for some scientists to use. So clock makers kept searching for still better resonators, which could produce many more, even smaller beads each second. They had known for some time that *atoms* vibrate, or resonate. Each kind of atom has its own set of vibration rates, or frequencies. The hydrogen atom, for instance, under certain conditions vibrates 420,405,752 times per second. The cesium atom has a *resonant frequency* at 9,192,631,770 cycles per second, or hertz. If the clock makers could find a way to use *atoms* as resonators and keep track of the vibrations, they would have a very fine clock indeed.

In 1949, the United States National Bureau of Standards announced the world's first atomic clock. In the years since then, National Bureau of Standards scientists have improved on this first clock with a series of newer models. Today's model, NBS 6, uses atoms of cesium, a soft, silvery metal, as its resonator. Electronic counters keep track of the 9,192,631,770 beads it produces every second of the day and night. It is so accurate that it will not lose or gain one second in 370,000 years—if it lasts that long! Located at the National Bureau of Standards laboratories in Boulder, Colorado, it is the frequency standard for the United States—the standard with which all other frequency standards in power companies, navigation control bases, scientific laboratories, TV and radio stations, telephone companies, electronic facilities, and many other places are *synchronized*. NBS 6, in turn, is compared regularly with the clocks of other nations by the International Time Bureau, in Paris, France.

How this remarkable clock works, and how scientists send its information about time interval and the time of day all over the world, 24 hours a day every day of the year, we shall find out in chapter 7. But first we must try to understand some of the problems created by having such an accurate time piece.

3

Every Second Counts

WITH ELECTRONIC CLOCKS and digital watches all around us, it's hard to imagine that just keeping track of what *day* it was once presented big problems. Trying to measure hours and minutes was even harder. But primitive peoples didn't care much about such small pieces of time. They got up with the sun, rested and ate their main meal around noon, and went to bed when darkness stopped their daytime activities.

What they needed most to know was when to plant their crops to take advantage of the best growing season. This date was so important that they spent enormous efforts over many years planning and laying out great stone "clocks" or calendars. By watching the progress of the sun and other heavenly bodies with respect to their stone markers, they could keep track of such important events as Midsummer Day, the longest day of the year. Stonehenge, in southern England, and elaborate stone structures in Central America and southern North America laid out by Maya and Aztec peoples, are examples of these earliest clocks. Built long before the people began to develop writing, they are found in all parts of the world. Today scientists are uncovering evidence that stone structures on our own western states, such as the Big Horn Medicine Wheel in Wyoming, were similar clocks, made by Indians.

Stonehenge is an ancient clock-calendar in southern England, dating from about 1848 B.C. If one stands at the center of the monument at dawn on Midsummer Day, he sees the sun rise directly over a certain stone marker. Recent computer studies have shown remarkable alignments of certain stones with various stars at different seasons of the year.

Because many of these sites were used for human burial and carvings on some of the stones are religious symbols, scientists once believed that the structure had only a religious purpose. No doubt there were religious rites connected with birth, death, planting and harvesting crops, and raising domestic animals. The movements of heavenly bodies, the endless succession of seasons, and time itself have always inspired a sense of mystery and awe. These great, orderly cycles continuing over thousands of years and so obviously beyond man's control *must* depend on some great, unseen power. So we find time, astronomy, and religion and worship closely related from the earliest history of man.

The earth swings around the sun, and the moon swings around the earth. The earth swings on its own axis. These movements can be easily observed and charted from almost any place on the earth, and the observations have always been useful in keeping track of time. It would have been much easier for astronomers and mathematicians if the periods of these great cycles had been neatly divisible into each other. But this is not the case. It takes the earth about 365¼ days to complete its swing around the sun. And the moon circles the earth about 13 times in 364 days.

This situation shows us another of the problems in measuring time. If one is measuring ribbon or counting liters of beans, he can start anywhere, stop whenever he likes, and continue when he finds it convenient. He can count fast or slowly, and if he comes up short a few beans in one measure, he simply adds some from his supply. If he has more ribbon than he wants in one piece, he can snip off the end and throw it away. But this is not the case in measuring time. Every single second of every minute, hour, day, and year must be accounted for, year after year after year.

So early calendar makers had some hard problems to work out. Farmers in the Tigris-Euphrates Valley, in Iraq, had made a calendar with 12 months in the year; each month was 29½ days, the average time between two new moons. This added up to 354 days in a year, 11 days short of the year we know. In a few years the farmers found that their planting time was out of step with the seasons. So they added extra days and months, at first on a random basis, but later at regular intervals over a 19-year cycle.

Egyptian astronomers were the first to see that the solar year was close to 365 days, with an extra day added every 4 years. But the astronomers could not persuade the rulers to add the extra day. So for 200 years, the calendar and the seasons slowly drifted out of phase. Then in 46 B.C., Julius Caesar proclaimed the 365-day year with an extra day every fourth year, or "leap" year.

But even this adjustment wasn't quite right. An extra day every four years amounts to an over-correction of 12 minutes each year. Some thousand years after Caesar established his calendar, this small yearly error added up to about 6 days. Easter and other important religious holidays were coming earlier and earlier each year.

By 1582 the error had become so great that Pope Gregory XIII changed the calendar and the rules for making it. First, years beginning a new century that are not divisible by 400 would not be leap years. For example, the year 2000 will be a leap year because it is divisible by 400. But the year 1900 was not. This change reduces the error to about one day in 3,300 years. Second, to bring the calendar back in step with the seasons, Pope Gregory removed 10 days from the year 1852 by a decree that October 4 of that year would be followed by October 15.

Adoption of the Gregorian calendar pretty well solved the problem of keeping the calendar in step with the seasons. But with all the problems and the adjustments in various calendars to try to solve them, it's easy to see why there are uncertainties about exactly when some historical events, such as the birth of Jesus, took place. Records do not agree, and in some cases it's impossible to know which calendar was the basis for the record. As we shall see later, many of the same *kinds* of problems confronted timekeepers who were concerned with much shorter cycles and pieces of time—minutes and seconds.

The candle clock was once the latest thing in time pieces.

Measuring time *interval* was, in fact, as important to some early professionals, scientists, and craftsmen as keeping track of the *date* was to others. In the 9th century, King Alfred the Great of England used candles to measure his working day. His candles, marked off in 3 sections, burned for 4 hours. Other candles burned for 12 hours.

Galileo and other scientists were observing falling objects and balls rolling down tilted planks, to try to understand the laws of motion. Metal workers wanted to heat and hammer out steel for swords that would neither bend because they were too soft nor break because they were too brittle. Measuring time interval was vital to these and other experiments and processes. But there were no suitable clocks.

Some craftsmen noticed that people tend to chant or sing a song over and over always at the same pace. And so elaborate rituals were developed, often including patterned marching and clapping or drum beating as well as chanting and singing, to time various processes. But of course this took a lot of man power—or woman power, since the chanters and marchers were usually young women.

The Greek and Roman parliaments had a different need for measuring time interval—one that exists today: They wanted to give each speaker an equal chance to be heard, but to limit the length of his speech. Their solution was a water clock, called a *clepsydra,* or "water thief." First built in Egypt, the water clock in its simplest form was an alabaster bowl, wide at the top and narrow at the bottom, with horizontal "hour" marks on the inside. The bowl was filled with water, which dripped out through a small hole in the bottom. Because of its design, it kept fairly uniform time.

The Chinese developed a clock with some features of later mechanical clocks. It was basically a water clock, for it used falling water to move a wheel with small cups arranged around its rim. As each cup filled with water it became heavy enough to trip a lever that moved the next cup into place.

Some Chinese water clocks were very elaborate.

The wheel thus revolved in steps, measuring the time. Many variations of the Chinese water clock were built in different parts of the world, and it had become so popular by the early 13th century that a special guild for its makers existed in Germany.

But besides the fact that it didn't keep very good time, the water clock often froze in winter. The sand clocks introduced in the 14th century avoided this problem; but because of the weight of the sand, they were limited to measuring very short intervals. The sand glass, or hourglass, was especially useful on ships. Sailors threw a log overboard, with a long rope fastened to it. As the rope played out into the water, they counted the knots that were tied into it at equal intervals, for a length of time determined by the sand glass. This gave them a crude estimate of the speed—or "knots"—at which the ship was moving.

The first mechanical clock was built probably sometime in the 14th century. It was powered by a weight attached to a cord wrapped around a cylinder. A fairly simple combination of balanced weights and notched wheels moved the clock's one hand around a dial. Since the clock lost or gained about 15 minutes a day, a minute hand would not have been very useful. During the 15th century, a spring replaced the weight in some clocks. But this didn't work well either because the spring became weaker as it unwound.

As long as the "works" of a clock depended mainly on the friction between parts, the strength of the driving weight or spring, and the skill of the craftsman who made it, time keeping was crude even at best, and no two clocks kept the same time. The need was for some sort of *resonator* that could be part of the clock itsef.

A swinging pendulum was just such a resonator. Although Galileo first realized that a pendulum could be the frequency-regulating device for a clock, it was the Dutch scientist Christian Huygens who built the first pendulum clock, in 1656. It was accurate within 10 seconds a day—a dramatic improvement over all previous clocks. Later clocks, with *two* pendulums, kept time to a few seconds in five years.

But there was one very serious limitation of the pendulum clock: it had to stand or hang solidly in one place. So it was useless on a tossing ship. And it was ships at sea that had the greatest need for an accurate, reliable clock to help them find their way safely across the vast, unmarked expanses of the great oceans.

4

How Time Tells Us Where in the World We Are

WHY IS A VERY ACCURATE, dependable clock so important on a ship? Because out on the open ocean, beyond sight of land, the ship's navigator has no way of knowing where he is unless he also knows what time it is—in Greenwich, England. Even today, with paths across the oceans well charted and with all the radios, computers, and fancy electronic instruments that help ships and airplanes stay on course, the one thing the navigator needs most to know is exactly what time it is at Greenwich. Small pleasure-boat navigators and private airplane pilots also need this information, correct within a second or less.

Many persons know that time is important to navigation, but few understand just why. Primitive peoples discovered long ago that the sun and stars could guide them on their travels—through the desert wastelands where sands were constantly shifting, and especially on water out of sight of familiar landmarks. Woodsmen, and city dwellers too, often observe the sun and shadows to "get their bearings" and keep track of which way they should be traveling. Some naturalists feel quite sure that migrating birds, flying sometimes over thousands of miles of open ocean, use the stars and sun to follow their courses.

Early explorers in the northern hemisphere were fortunate in having the North Star, or "pole star," that appears to be suspended in the northern sky at night. Unlike the other stars, it does not rotate or change position with respect to the earth. These early adventurers also noticed that as they traveled northward, the North Star appeared higher and higher in the sky. At the North Pole it is directly overhead.

By measuring how far the North Star was above the horizon, then, a navigator could figure out his distance from the North Pole—and also from the equator. An instrument called a *sextant* helped him measure the angle formed by the North Star, his ship, and the horizon. The measurement is indicated in *degrees of latitude,* ranging from 0 degrees at the equator to 90 degrees at the North Pole.

So the navigator could tell quite well just where he was on a north-south line. But measuring distance east and west was a different and much harder problem because of the earth's spin. The problem, however, also supplies the key to its solution.

For measurements in the east-west direction, the earth's surface has been divided into lines of *longitude,* or *meridians.* All meridians meet at the North and South Poles. The imaginary line that runs through Greenwich, England, has been labeled the zero meridian; and longitude is measured east and west from this meridian to the point where the measurements meet at 180 degrees, on the opposite side of the earth from the zero meridian.

Now, at any point on the earth the sun travels across the sky from east to west at the rate of 15 degrees in one hour, or one degree in 4 minutes. So if a navigator has a very accurate *clock* on his ship—one he can depend on to tell him very

A sextant helps a navigator find the location of his ship with respect to heavenly bodies.

precisely what time it is at Greenwich or the zero meridian—he can easily figure out how far east or west of the zero meridian he is. He simply gets the time *where he is* from the sun. Then for every 4 minutes difference between the time in Greenwich and the time where he is, he is one degree of longitude away from the zero meridian.

At night the navigator can find his location by measuring the positions of two stars with respect to his ship.

The North Star, as we've said, seems from his ship to hold a steady position. But the spin of the earth makes the other stars appear to move in circular paths around the North Star. The navigator has charts that tell him the location of the stars as they appear from any place on the earth, at all times and in every season—according to the *time* in *Greenwich*. So by using his sextant to measure the locations of the stars, and his clock to find the time in Greenwich, he refers to his charts to find out exactly where he is. An error of just one second on his clock, however, can mean an error of ½ kilometer in his location, and send the ship far off its course.

The method of charting and holding the ship's course is very neat and fairly simple. A Flemish astronomer named Gemma Frisius had suggested using a timepiece to chart longitude as early as 1530. The problem was that until a little over 200 years ago, no one was able to come up with a dependable clock that would keep accurate time on board a ship. The pendulum clock that had so greatly improved time keeping on land was useless on a rolling, pitching ship. So although navigators could keep track of their north-south location very well, they had to go on for hundreds of years depending on "dead reckoning" for their east-west location. And although many developed an uncanny ability to guess the ship's position, their reckoning was all too often dead wrong.

During the centuries of exploration and colonization of the New World, the need for better navigation aids became pressing. Ship building improved, and larger, stronger vessels made ocean trade, as well as ocean warfare, more and more important. But it was not unusual for shiploads of settlers or cargo ships carrying slaves or priceless merchandise to be lost at sea. Driven off course by storms or changing course to avoid bad weather or outrun a pirate ship, they had no way to tell their longitude or which way to head for a safe harbor. The navigator could not tell how far they were from their goal. The ship might smash to pieces on a rocky shore, or land some place far from its intended destination.

By the beginning of the 18th century the need for better navigation instruments—specifically for a better clock—had become desperate. In 1713, the British government offered an award of £ 20,000 to anyone who could build a clock, or *chronometer,* that could be used on a ship and that would serve to determine longitude at sea to within 1/2 degree.

Among the many craftsmen who hoped to win this large prize was an English clock maker named John Harrison. He spent more than 40 years trying to meet the tough specifications. Each model he made seemed a little more promising as he found ways to overcome the many problems. Springs and ingenious "escapements" had to take the place of the simpler and more dependable pendulum. Temperature changes caused expansion and contraction of delicate metal parts. The old problem of friction between moving parts had to be reduced as much as possible. Salt spray corroded everything aboard ship.

When Harrison finally came up with a chronometer that he thought was nearly perfect, the men who offered the prize were so afraid it might be lost at sea that they put off testing it until Harrison had built a second one exactly like it so that there would be a pattern. At last, in 1761 Harrison's son William was sent on a voyage to Jamaica to test the instrument. The ship ran into a severe tropical storm that lasted for days and drove the ship far off its course. The navigator insisted that they were in a location much different from that calculated by William with the help of the chronometer. But the captain decided to rely on the new instrument, and the ship sailed triumphantly into the harbor at Jamaica just as William had predicted. The chronometer lost less than one minute over a period of many months, and had made it possible for William to determine his longitude at sea within less than 1/3 degree. John Harrison claimed the £20,000 prize. Part of it had been given to him earlier for less

This modern chronometer with its quartz-crystal oscillator looks much like Harrison's instrument.

perfect models of his clock, and he received the rest in varying amounts over the next two years—just three years before he died.

For the next half century a chronometer much like Harrison's—each one built individually by a skilled crafts-man—was one of the most important and treasured pieces of equipment on any ship. It needed very careful tending, and the one whose duty it was to care for it had a serious responsibility as well as a position of honor. When the ship was in port, the captain often took the chronometer ashore for safe keeping.

Today the navigation system on a ship or in an airplane includes radio equipment, computers, sonar, radar, and gyroscope compasses. There may be a hundred or more quartz-crystal wristwatches among the crew, any one of which is even more reliable and accurate than John Harrison's instrument. But the chronometer, built on the same basic principles as Harrison's instrument, is still the heart of the ship's or plane's or submarine's navigation system. And knowing exactly what time it is at Greenwich is as necessary to the navigator today as it was centuries ago. Now, though, his modern equipment can give him the time accurate to a microsecond, so that he can locate his position within a few hundred meters. We shall discuss these systems more fully in Chapter 15.

5

The Growing Needs for Better Clocks

ON AN AFTERNOON shopping trip you may pick up a roll of 8mm film for your dad and the 35 ml can of frozen orange juice your mother needs. At the shoe store you tell the clerk you want to look at size 8½ tennis shoes, and then go to the ski shop to see whether the 207 cm skis your instructor said you should have are ready with the new bindings. *Standards,* today, are all around us. The 50-meter relay covers the same distance in Los Angeles as it does in Chicago—or in Munich.

Every kind of measurement has its official, recognized, basic unit. The basic unit for measuring temperature is the degree—Farenheit or Celsius, depending on which *scale* you are using. For weight it is the kilogram, and for length, the meter. These and many other basic units are still rather new and strange in the United States, except among scientists and a few others. They are probably more familiar to you than to your parents, for until 1975 everyday America continued to use the English units and scales that had been the bases for measurements for hundreds of years. But the United States government—especially the Department of Commerce—has made strong efforts to get Americans to give up the old scales

27

based on inches and ounces and adopt the simpler *metric* scales used by every other advanced nation in the world.

The basic unit for measuring time interval is the second. Sixty seconds make one minute, and 60 minutes make one hour. Fractions of seconds are measured in decimals—tenths, hundredths, thousandths or *milliseconds,* millionths or *microseconds,* and billionths or *nanoseconds.* The definition—and therefore the length—of the second itself has been changed slightly from time to time according to the *scale* by which it is measured. But as early as 1820, French scientists defined the second as 1/86,400 of the mean solar day, since there are 86,400 seconds in 24 hours. With only very small changes, this definition was accepted throughout the world until quite recently, in 1967, when modern communication and scientific needs made it necessary to make a basic change, as we shall see in the next chapter.

Many of the standards we know today resulted from improved communication and transportation sytsems. As long as people lived spread out on farms or in mining camps—or in villages with little communication with people in the next village—standards were not too important. Buyer and seller could agree on the price of a "box" of apples or a "load" of hay they were inspecting. A race might be run "from the oak tree behind the post office to Mr. Smith's fence corner." But when better roads, and especially telephone and telegraph lines made it possible to communicate with persons some distance away, there had to be some specific way to describe a box of apples or a load of hay. A woman who telephoned to another for a recipe needed to know just how much flour and sugar to put into the cake. So uniform weighing scales and measuring utensils became common.

The time of day—or *date*—was important to running the community affairs, but no one cared much what time it was somewhere else. So each city, town, and village had its own time, based roughly on sun time at its own location. A local jeweler's best clock, or the outdoor clock at the courthouse or bank served as the standard. Or sometimes it was the local factory whistle that blew at 7:00 in the morning, 12:00 noon, and 1:00 and 4:00 in the afternoon to tell the workers when to start and stop work each day, that governed school and business openings and closings as well. The time of day in the next town might be fifteen minutes or half an hour different, but that didn't matter.

It was the growth of the railroads that began to make the time of day matter a great deal. Railroad men traveling with the trains couldn't be forever changing their watches to match the many local times shown by the cities they passed through. And there had to be some kind of dependable scheduling for arrival and departure at each station.

So for a number of years railroads kept their own time. But each one had a different time, based on the time in the major city through which it passed. At the railroad station in any city, one found a row of clocks on the wall, one for each line that served the city, and each telling the time according to its own standard. Long lines that crossed several states used different bases in different places. A passenger fortunate enough to own a watch might have to change it 20 times if he wanted it to be "right" as he traveled across the nation. It's a wonder that anyone ever caught the train he intended to, and that there weren't many more train wrecks than there were.

Clearly, something had to be done to establish order. The need was for some sort of time standard, or *standard time*. The need was greatest in the United States because of the wide distances spanned east and west. Railroads in England uniformly adopted Greenwich time, and those in Europe used Paris time. The fact that these two didn't agree with each other mattered very little, since there was no connecting link between them. But of course American and Canadian railroads could not be expected to adopt Washington time throughout the land.

The idea of dividing the nation—and the world—into

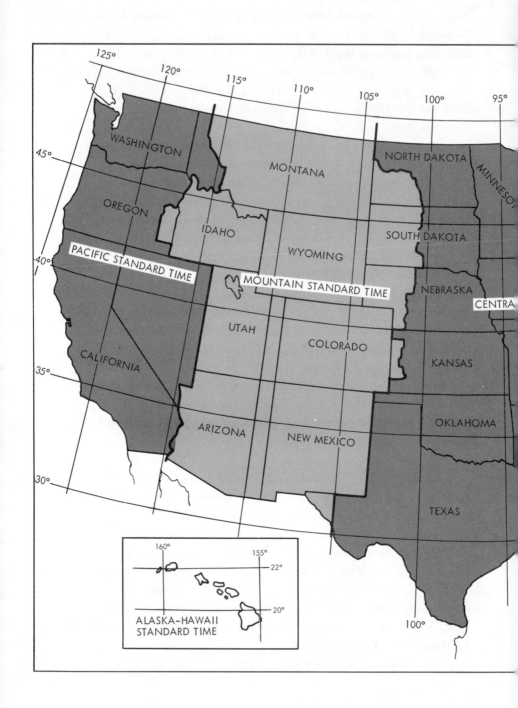

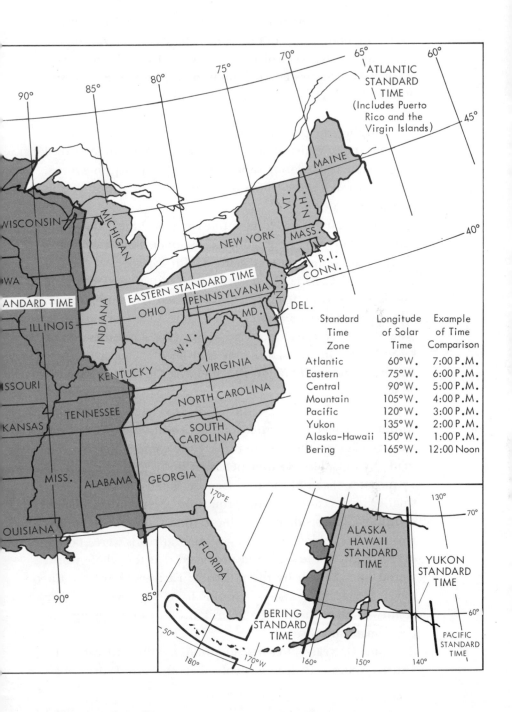

ATLANTIC
STANDARD
TIME
(Includes Puerto
Rico and the
Virgin Islands)

EASTERN STANDARD TIME

ANDARD TIME

Standard Time Zone	Longitude of Solar Time	Example of Time Comparison
Atlantic	60°W.	7:00 P.M.
Eastern	75°W.	6:00 P.M.
Central	90°W.	5:00 P.M.
Mountain	105°W.	4:00 P.M.
Pacific	120°W.	3:00 P.M.
Yukon	135°W.	2:00 P.M.
Alaska-Hawaii	150°W.	1:00 P.M.
Bering	165°W.	12:00 Noon

ALASKA HAWAII STANDARD TIME

YUKON STANDARD TIME

BERING STANDARD TIME

PACIFIC STANDARD TIME

time zones was discussed in government and commercial circles in both Europe and the United States. One of the most eager promoters of the idea was a Connecticut school teacher, Dr. Charles Dowd, who wanted to make travel easier and railroad crossings safer. He talked to railroad and government officials, and anyone else who would listen to him, urging support for his plan. Since the continental United States spans about 60 degrees of longitude, Dowd suggested that the nation should be divided into 4 time zones, each 15 degrees wide, which is the distance the sun travels in one hour.

After years of prodding from Dr. Dowd and others, the railroads and some cities adopted his plan. Four time zones covered the United States, with a fifth for the easternmost provinces of Canada. At noon on November 18, 1883, time "stood still" while telegraph lines carried the United States Naval Observatory's master clock time to major cities. In New York City, the clocks were stopped for 3 minutes and 58 seconds—the amount local time had been ahead of the new Eastern Standard Time. In Philadelphia the change was only 36 seconds.

Looking back, we would think that such a good plan would have met with immediate approval, but this was not true. In Chicago the citizens objected to conforming to a rule made by outsiders. So they persisted in keeping their clocks 17 minutes ahead of the new Central Standard Time. Many smaller cities and towns, proud of their independence, continued to use their own local time. Newspaper editors wrote articles objecting to the way the railroads were "taking over the job of the sun" and predicted that the whole world would soon be run by the railroads. Farmers were sure the new "unnatural" system would have all sorts of dire results—from reduced egg and milk production to changes in the climate and weather.

But 25 years after its adoption by the railroads, the United States was deeply involved in the first World War.

Tanks, guns, food, supplies, and thousands of troops had to be moved quickly and efficiently all over the country. On March 19, 1918, the United States Congress passed the "Standard Time Act," authorizing the Interstate Commerce Commission to establish standard time zones within the United States. The act also established "daylight saving time," to save fuel in a country at war. The rules for daylight saving time have been changed several times in recent years. The main need is to keep them uniform, so that wherever the shift is made, it's made at the same time.

Today the United States is divided into 8 time zones, including Alaska and Hawaii. The boundaries between zones are not all straight, but zigzag here and there to accommodate cities near the boundary and place them in the same time zone as the larger cities with which they naturally trade or do business. Out on the oceans, the time zone boundaries have been adjusted for island groups, for similar reasons.

The world is divided into 24 standard time zones, with the zero zone centered on a line running north and south through Greenwich, England. The time zones east of Greenwich have time later than Greenwich time, and those to the west have earlier times—one hour difference for each zone.

With this system a traveler who crosses the International Date Line, 180 degrees around the world from Greenwich, gains a day if he is traveling east to west. If he is going in the opposite direction, he loses a day.

By the early 1900s "railroad time" was very generally available and was often used as the local source for the "right" time. The time of day was sent throughout the system by telegraph, and fast trains were often scheduled just minutes apart on the same track. Railroad crews took great pride in bringing their train in right on the minute. Every "railroader," from the station manager and dispatcher to the engineer, conductor, and brakeman—and even the track repair crew with their handcar—had to know the time, often to the part of a minute.

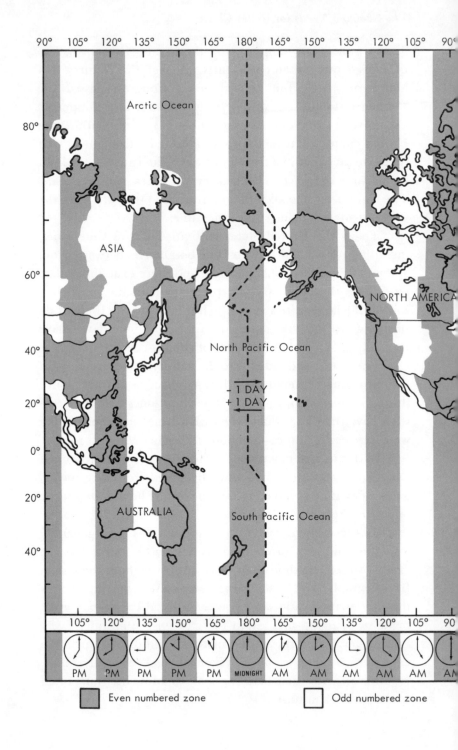

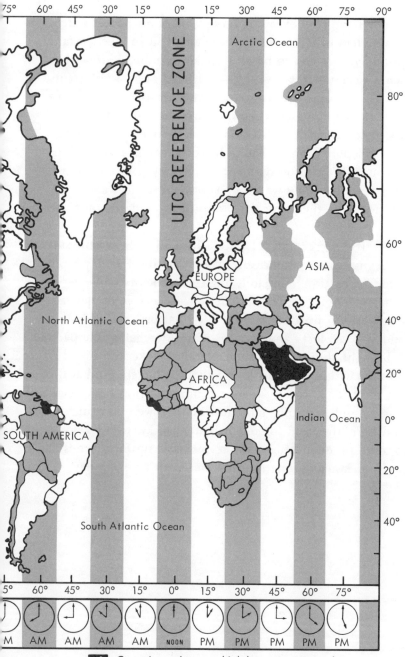

Countries and areas which have not accepted the zone system, or where time differs more or less than one half hour from adjacent zones.

A railroad employee was very proud of his pocket watch, which he had to buy himself and which had to meet certain requirements. It had to be of a certain minimum size—wrist watches were not yet invented—and have 21 "jewels" in its bearings to reduce the friction between moving parts and thus insure its accuracy. Each morning the railroader had to check his watch with the signal coming in over the telegraph wire or telephone, and it had to be correct at all times within 30 seconds. Watches were checked on the job by inspectors, who appeared without warning. Today electronic watches are permitted, but the same strict accuracy standards must be kept.

Electric power and telephone lines as well as railroads had begun to tie Americans closer together, and people became more time conscious than ever before. The wrist watch and dainty ladies' watches, masterpieces of delicate craftsmanship, became fashionable. Factory assembly-line production of clocks and watches with interchangeable parts had brought the price of time pieces that kept reasonably good time within the reach of many more buyers. And at the many electric generating plants, communication equipment factories, and scientific laboratories that were starting up everywhere, the need for much better clocks, whose resonators could produce much smaller, precisely uniform beads of time for measuring *time interval*, was growing every day.

6

Flaws in the Earth-Sun Clock

IN CHAPTER 2 we said that every clock has to have a resonator and a source of energy to keep it running. But the less the energy source "disturbs" the "natural frequency" of the resonator, the better. A perfect resonator would be one that, with just one "push" by its energy source, would run forever. But of course this is not possible; because of friction, everything "runs down." A swinging pendulum comes to a stop unless a spring or weight keeps supplying energy.

Scientists often describe a resonator in terms of its "Quality Factor," or "Q." Q is the number of swings or vibrations the resonator makes before the energy it gets from one push of its energy source is nearly gone. An average mechanical watch might have a Q of 1000. A clock with a quartz-crystal resonator has a Q of 100,000 to 1,000,000, depending on the quality of the crystal. Today's atomic clocks have Q's in the millions.

One important advantage of a high-Q resonator is that unless it runs at or very near its natural frequency, it won't run at all. So if it's running, we can be sure of its *stability* and *accuracy*. To understand these terms let's think about a machine that's filling pop bottles. Hour after hour, whenever it's running, the machine is supposed to put exactly one liter of pop into each bottle. As we watch this machine, we see that it does put almost exactly the *same amount* of pop into each bottle—within 0.2 liter perhaps. So we say that the machine has good *stability*. But then we also notice that each bottle is only ¾ full. So we say that although it has good stability, the machine has poor *accuracy*.

This situation might be the reverse. A different machine might be putting a few extra milliliters of pop into some of the bottles and leaving others short. The overall *average*, however, is one liter per bottle. This machine has good accuracy, on the average, but poor stability.

A clock that gains or loses a predictable number of seconds or microseconds each day may have high stability but poor accuracy. A clock that runs "fast" under some conditions and "slow" under others may have good accuracy if it "comes out even" at the end of a day or month, but poor stability. Good accuracy and good stability do not necessarily go together. Clock makers and scientists, of course, want clocks that have resonators with high scores in both accuracy and stability.

As scientists produced better and better clocks—with higher and higher Q's—they were able to prove some things about earth movements that they had suspected for a long time. They found that although the earth-sun clock had great overall accuracy, its stability, by more precise standards, is really not very good. Running sometimes "fast" and sometimes "slow," it does not produce a steady rhythm.

There are several reasons for these irregularities. The axis of the earth's rotation is not at right angles to the plane of its orbit, but tilts at an angle of about 23½ degrees. And the earth's orbit around the sun is not a perfect circle, but a slight oval or *ellipse*. So the earth travels faster when it is nearer the sun—in winter, in the northern hemisphere—than it does when it is farther away—in summer. Because of these

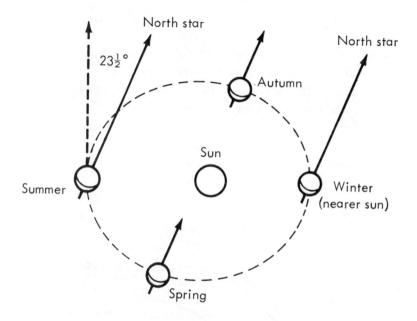

North star

$23\frac{1}{2}°$

Autumn

North star

Sun

Summer

Winter
(nearer sun)

Spring

Motion of Earth Around Sun

two factors, the time of day as measured by a sundial can differ from the "norm" by as much as 15 minutes a day in February and November. These differences are predictable; so they don't cause too many problems for timekeepers. Various time scales have been developed that "smooth out" these irregularities quite well.

But there are other variations that are *not* predictable. The spin of the earth on its own axis is not perfectly regular, and it even wobbles a little on its axis, with the North and South Poles wandering around by a few meters from year to year. Scientists are not sure just why these variations exist. Probably such things as shifts in the molten core of the earth, and even the piling up of snow on the mountains in winter and the melting of this snow in summer, are responsible.

The atomic clocks developed in the 1950s were one hundred thousand times more stable than the earth-sun clock. With such a standard, scientists could study and measure variations in time caused by the irregular movements of the earth more precisely than ever before. Their measurements proved that several thories that had been advanced by various scientists were true.

They found, for instance, that as they suspected, the earth's spin slows down a few milliseconds in the spring—in the northern hemisphere—and speeds up in the fall. Precise

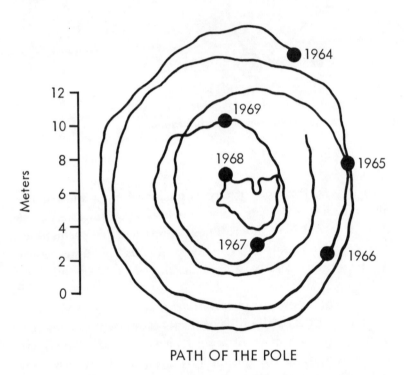

PATH OF THE POLE

Earth wobbles as it rotates about its axis, so the poles seem to migrate. The plot above shows the apparent North Pole wandering over a range of about 15 meters. The actual coordinates of the pole must be used to correct mean solar time for the use of sailors and aviators navigating by the sun and stars—a change of about 0.05 second.

measurements showed that the wobbling axis causes variations of as much as 30 milliseconds from year to year. They
also showed that the earth is gradually slowing down. The
length of the day is about 16 milliseconds longer now than it
was 1,000 years ago.

So who is going to worry about a few irregularities that
even added together, amount to less than one second in a
whole year! Most of us would agree that it matters not at all,
and that scientists who spend their time studying such things
and trying to do something about such trivial differences are
wasting their effort and the taxpayers' money. That is—until
we realize that if we are to have telephones, radios, TV sets,
computers, and hundreds of other electric and electronic
devices that we all take for granted, the problems created by
these tiny irregularities in solar time *had* to be solved.

The scientific and industrial users of time and time
technology *must* have extremely *small, uniform* beads of *time
interval,* such as those measured out by an atomic clock;
atomic clocks were, in fact, developed in answer to this need.
Navigators of ships and airplanes need precise information
about *date,* as shown by solar time, to avoid accidents and
keep their craft on course. Somehow a way had to be found to
satisfy both kinds of time users—to make the brown shoe and
the black shoe we mentioned in Chapter 1 serve as a pair.

Scientists and government laboratories around the world
have spent many thousands of hours and large sums of money
to find a system that will work. By the mid-1950s several
scientific laboratories had begun to use the new atomic time
scale. Another time scale, called *Coordinated Universal
Time* or *UTC,* was the official scale throughout the world.
Based on solar time, it modified actual solar time in several
minor ways to "correct" its irregularities and make the rate,
or size of the time beads, as uniform as possible.

Various schemes were tried to keep UTC and solar
time in step. UTC clocks were controlled by atomic clocks, to
keep the time beads uniform. But then because of the

unpredictable variations in the earth-sun clock, the *size* of the time beads had to be adjusted very slightly at the beginning of each new year. This actually changed the length of the second, but kept the same *number* of seconds in each year. On the basis of the year that had just passed, scientists tried to predict exactly how long the next year would be, and to adjust their official clocks to run at the proper rate. But then the year never came out just as predicted.

This practice, which began in 1958 and was used for the next 14 years, resulted in "rubber" seconds that could be stretched or shrunk a little as the predictions required. The system proved to be a costly nuisance. Each year, very fine and precise clocks all over the world had to be adjusted to run at a slightly different rate. There were also tenths of seconds added to or subtracted from the date every once in a while, to keep the system within 1/10 second of sun time. The effect was similar to what we might expect if each year the length of the centimeter were changed slightly, and all meter sticks—made of rubber, of course—had to be stretched or shrunk to fit the "centimeter of the year."

One fact was very clear: either the rubber seconds would have to continue indefinitely, or there would have to be some provision to vary the number of seconds in a year. There was certainly no way to make the earth turn in exactly the same way every year; and every second, you'll recall, must be accounted for.

The rubber seconds were not popular with anyone. So in 1967, the old official definition of the second, based on astronomical measurements, was dropped, and the new atomic second was adopted worldwide. Since then the official, standard second has been defined as 9,192,631,770 oscillations—or vibrations—of the cesium atom.

This change provided a good, firm second that can be measured precisely within a few billionths of a second. And scientists and others with the necessary equipment can make the measurement easily in less than a minute. Electronic devices that are part of the atomic clock count the vibrations of the atoms very precisely and display the count in the same way that a pendulum clock counts and displays the swings of its pendulum.

Of course, this new definition of the second has nothing to do with the earth rotation or the sun and stars. So atomic time and earth time continue to get out of step. To solve this problem, the "leap second" was invented in 1972. The leap second is much like the leap year, when an extra day is added to February every fourth year to keep the year in step with the movement of the earth. A leap second is added—or possibly subtracted—when the irregular movement of the earth makes this necessary. The rule is that atomic time will always be within 0.9 second of solar time. The leap second is added to or subtracted from the last minute of the year, in December. Or an adjustment may be made in the last minute of June. The adjusted minute thus becomes either 61 or 59 seconds long. The International Time Bureau, in Paris, France, tells timekeepers throughout the world when the change is to be made.

In 1972, which was a leap year, two leap seconds were added, thus making it the "longest" year in modern times. One leap second has been added each year since then, in December.

This present system for keeping track of time, although a big improvement that works quite well, is still imperfect; and scientists are always searching for ways to make it better—and to build even better clocks.

7

Inside the Atomic Clock

IF YOU WERE to join a group of visitors at the National
Bureau of Standards laboratories in Boulder, Colorado, one
of the high points of the tour would be the famous "atomic
clock"—or more properly, the national frequency standard,
NBS 6. And you would probably be disappointed. Your first
response might be, "That's a *clock?* Where's the *time?*"

Certainly it looks like no other clock you've ever seen
before. A stainless steel pipe about 6 meters long and perhaps
40 centimeters in diameter, it looks more like some exotic
piece of plumbing than a clock. A few glass-covered port
holes stare vacantly at the white, sterile-looking room, where
nothing at all seems to be going on. The tube just lies there
in its rack, like a section of very fancy sewer pipe.

You would not be permitted in the room itself, but
would view the clock through a window from the hallway. If
you hadn't already noticed it and asked the question, the
guide would point out the one doorway to the room, with its
solid copper casing. He would tell you that the entire room,
behind the plaster walls and ceiling and beneath the floor, is
also lined with copper. This is to stop any radio or electric
influence from outside that might disturb the clock.
Temperature and humidity are carefully controlled.

National Bureau of Standards Atomic Clock.

Under the dull gray "bench" on which the long tube lies, you'd notice a few steel cases with knobs and gauges on the front. But even these don't tell the time. Somehow, the whole thing is a kind of letdown. Nothing moves, nothing ticks or buzzes. There are no flashing lights or swinging pendulums.

Across the narrow hallway, through another window you would see a neat display of several clock faces—small, business-like, easy-to-read dials, all alike and similar to the faces of hundreds of common alarm clocks you've seen in drug or department stores. The second hands on all of the clocks jerk forward regularly, in step like marching soldiers. There are also two digital clocks, one labeled UTC and the other

Control Room, across the hall from the NBS Atomic Clock.

Atomic Time, and you are told that all of these clocks show the output of NBS 6, across the hall. The guide may explain that the system is really a "paper clock," meaning that the output of NBS 6 is read, counted, and displayed electronically by a computer, and then the clocks which you see are simply kept running in step with NBS 6. These clocks monitor each other, and the output of each one is also recorded by a computer.

If you happen to have a quartz-crystal watch, you have a wonderful opportunity here to set it very accurately. But after checking your watch and perhaps resetting it, you become restless. There's really nothing very interesting in watching a group of identical clocks doing exactly what you'd expect them to do.

It's what we *cannot* see, however, that makes atomic clocks exciting. After all, we know that no one can actually see atoms, even with a microscope. We know what atoms must be like only because we can observe how they behave in different situations. So let's look at what goes on inside an atomic clock—specifically a *cesium-beam tube* like NBS 6. Other kinds of atoms—rubidium and hydrogen—as well as ammonia molecules have also been used as resonators for atomic clocks which have somewhat different structures. But it's the cesium atom that now gives us the world-wide basis for measuring the standard second.

At room temperature, cesium is a soft, silvery metal that looks somewhat like mercury. The core or *nucleus* of a cesium atom is surrounded by a swarm of electrons; the one electron that's farthest out from the nucleus is in an orbit all by itself. This electron spins on its axis, creating a *magnetic field;* we could think of it as a tiny magnet. The nucleus of

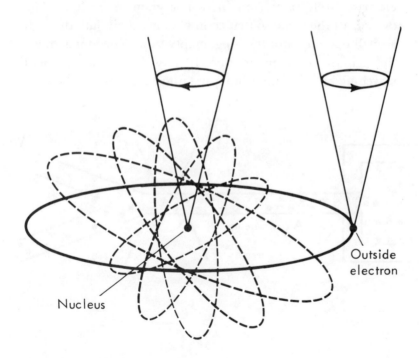

Outside electron

Nucleus

Cesium atom

the atom also spins, producing another tiny magnet. Each magnet "feels" the force of the other.

These two magnets are like spinning tops. If both have their "north" or + poles pointing in the same direction, the tops will be spinning in the same direction. But if the north poles point in *opposite* directions, the two tops will be spinning in opposite directions. So we have atoms in two different *energy states*. Atoms in one state are ready to *give off* energy, and those in the other state are ready to *absorb* energy. At a *frequency* of exactly 9,192,631,770 Hz, the outside, spinning electron of the cesium atom can "flip over," or reverse its north and south poles, either absorbing or giving off energy as it does so.

How can this activity be made to provide a resonator for a clock? The illustration shows us. On the left is a small electric "oven" that contains a few grams of cesium. The oven heats the metal so that cesium atoms "boil out" through a small opening into the long, empty tube. The atoms travel through the tube in an orderly way, like members of a marching band, without bumping into each other or the sides

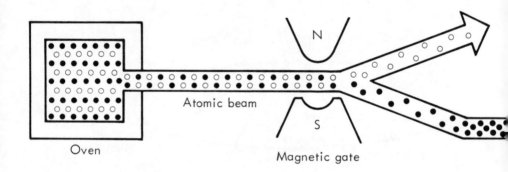

Oven

Atomic beam

N

S

Magnetic gate

O Atoms that pass through to detector

● Atoms spinning in wrong direction that are shunted away

of the tube. Near the oven they come to a "gate" which is a special kind of magnetic field that separates the atoms into two streams. Those with their outside electron spinning in one direction are allowed to go on through the tube. Those with the electron spinning in the other direction are shunted away.

The atoms that are allowed to go on then move into a part of the tube where they are "bathed" in a *radio signal* very near to 9,192,631,770 Hz—the *resonant frequency* of the atoms themselves. This causes large numbers of the atoms to "flip over" or change their energy state. The nearer the radio frequency is to the resonant frequency of the atoms, the larger the number that will flip over.

The atoms then pass through another magnetic gate that again separates them into two streams. Those atoms that have flipped over and changed energy state during their radio bath—those that have reached the resonant frequency—are allowed to go on to a *detector* at the end of the tube. Those that did not change cannot pass.

The detector produces a signal that is strongest when the greatest number of atoms are reaching it. It tells a quartz-

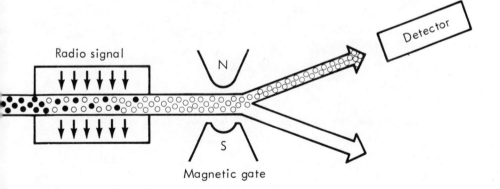

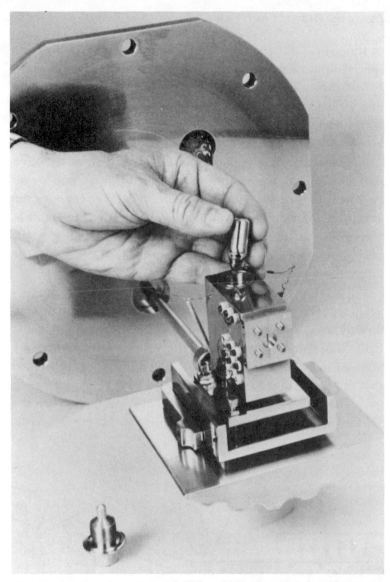

A capsule of cesium provides the atoms for a cesium-beam clock.

An electric oven heats the liquid cesium so that atoms boil out into the tube.

A detector tells the resonator when the greatest number of atoms are passing through.

crystal resonator when this is happening. And the crystal resonator, in turn, keeps the *frequency* of the radio bath just where it will cause the greatest number of atoms to flip over. This whole *feedback* system operates in a continuous cycle, automatically. The process is much like carefully tuning in a radio so that the listener hears the loudest and clearest signal. When this happens, the receiver is exactly "on frequency" with the signal that is sent. In the case of the atomic clock, the cesium atoms thus provide the broadcast signal and maintain the frequency at 9,102,631,770 Hz—which, we recall, is the standard definition of the second. It does this with such precision that the "clock" will not vary by one second in 370,000 years—or more importantly, by more than 3 millionths of a second in a year.

But how can the output of this incredibly accurate clock reach people who are not right where it is? In another part of the building, perhaps, or a hundred kilometers away. Or halfway around the world on a ship at sea. Or maybe thousands or millions of kilometers away in outer space.

The answer is by radio or satellite communication. As we mentioned in Chapter 5, time signals were sent by telegraph before the middle of the 19th century to help the railroads control the time schedules of their trains. With the development of radio in the early part of the 20th century, both public radio stations and scientific organizations broadcast time signals. At first these were sent only at regularly scheduled times—much as we receive them on commercial radio today.

The United States Naval Observatory, the National Bureau of Standards, and the International Time Bureau were pioneers in broadcasting time information. Today these organizations and others broadcast very precise time and frequency information in a variety of ways. The most far-reaching and widely used services come from shortwave radio stations such as the NBS stations WWV and WWVH. WWV is located at Fort Collins, Colorado, about 80 kilometers

Radio station WWV broadcasts precise time information continuously.

north of the NBS laboratories in Boulder. The sister station, WWVH, is at Kekaha, Kauai, the northernmost of the Hawaiian Islands. Both broadcast continuously the time and frequency information generated by NBS 6. Because such things as storms, atmospheric conditions, and sun spot activity often make it difficult for listeners in some areas to tune in the signal, both stations broadcast at several different frequencies at once in the 2.5 to 30 MHz range. Listeners with shortwave receivers in nearly all parts of the world can usually tune in one station or the other, on at least one of these frequencies, at some time of the day or night.

In addition to a voice announcement every minute that tells the time of day in Greenwich, England—or UTC—there is a "tick" each second that sounds like the ticking of a metronome or a grandfather clock. To avoid confusion for listeners where the WWV and WWVH broadcasts overlap, a man's voice tells the time from Colorado; and a woman's voice from Hawaii.

The "programs" of the two stations are very much the same. Tucked in between the time announcements and along with the ticks are two standard audio frequencies at 500 and 600 Hz, as well as occasional "Geoalerts" about current solar and geophysical events, and marine storm warnings. For

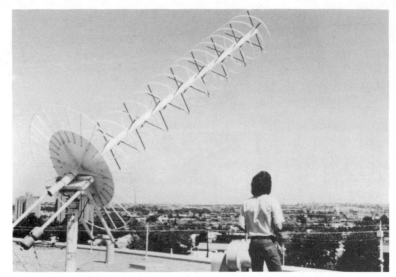

Antenna atop U.S. Bureau of Standards laboratories in Boulder, Colorado, monitors broadcasts from GOES.

A clock displays time received via satellite.

those who wish to tune a musical instrument and others with a need for such information, a standard 440 Hz tone, the musical note "A" above middle C, is broadcast once each hour. Both stations also carry a time code that can be picked up by persons with the necessary equipment; another NBS radio station, WWVB, which is located at Fort Collins, broadcasts only a time code.

In 1974, the National Bureau of Standards began sending a time code via Geostationary Operational Environmental Satellites (GOES) maintained by the National Oceanic and Atmospheric Administration. The code is available to most of the western hemisphere from satellites in orbit 36,000 kilometers above the equator. They travel about 11,000 kilometers per hour and remain continuously above the same spot on earth. Ham radio operators and anyone else interested in knowing more about the National Bureau of Standards stations can write to the Time and Frequency Division, National Bureau of Standards, Boulder, Colorado 80302.

The National Bureau of Standards shortwave stations are not the only sources of precise time information by radio. More than 30 radio stations throughout the world broadcast standard time and frequency signals, and there are also radio navigation systems that can be used as sources for time and frequency information. We shall discuss some of these in Chapter 15. But first let's follow the story of the regular broadcasts a little further.

Although radio signals travel at nearly the speed of light, it takes a very small fraction of a second for information to travel from the radio station to the listener. When the listener hears that the time is ten o'clock, it is really a tiny fraction of a second after ten. And the fraction is different for listeners in different parts of the world, and even different at different seasons and times of day in the same place. This is because the radio signal goes bouncing along its zigzag path between the earth and the *ionosphere,* a layer of the upper

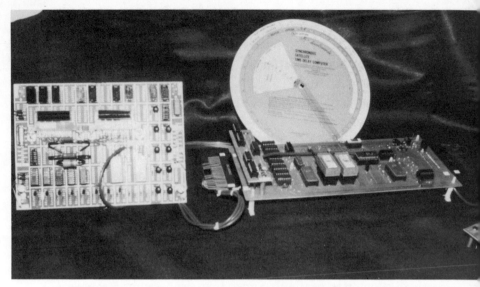

A microprocessor decodes time and position information from a GOES.

atmosphere that acts like a mirror to reflect radio waves. The shape and height of the ionosphere are constantly changing in unpredictable ways. So although the accuracy of the time *broadcast* is almost as precise as NBS 6 itself, it is hard for listeners to *receive* the signal with an accuracy better than one one-thousandth of a second.

For about 98 percent of the everyday users of time information, this degree of accuracy is fine. But for others, such as scientific laboratories, computer systems, and makers of electronic equipment or other atomic clocks, for instance, even this small, unpredictable irregularity causes problems.

The need for greater accuracy has led to several clever schemes for solving the problems. One of the commonest ways is to measure the *path delay* by sending a signal or pulse from a master clock, at a known instant, to the location of the clock to be synchronized. As soon as the signal reaches the location, it is sent back to the master clock. Then its arrival time back at the master clock is noted. Subtracting the transmission time from the arrival time tells how long the round trip took, and dividing this figure by two tells the path delay time of the one-way trip. Sometimes a satellite is used to relay signals back and forth between the locations of the clocks to be synchronized.

Another method for comparing clocks at widely separated locations is simply to carry the time, physically, from one location to the other. Scientists or technicians who wish to synchronize their clock with NBS 6, for instance, may take it to the NBS laboratories in Boulder to have it *calibrated*. Calibration of clocks and frequency standards is an important service of the laboratory.

Most of these "clocks" that are brought in periodically are *portable* cesium-beam standards. This much smaller

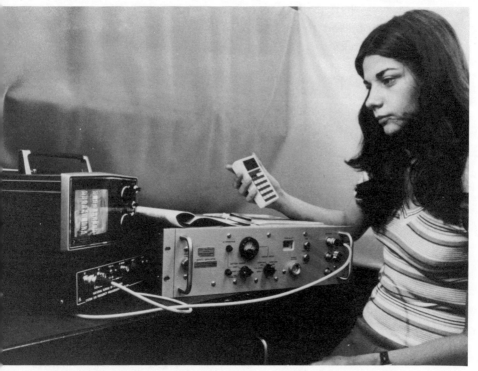

Calibrating a frequency standard.

version of the national frequency standard is packaged in a steel case about the size of a large piece of luggage. It can be carried by one or two men who tend the clock carefully on its travels—usually on a commercial airplane, as a passenger! This is so that the clock can be plugged into the plane's electrical system as it travels. The clock has a battery pack that supplies energy to keep it running when it is not where it can be plugged into some other power supply.

Four times a year, NBS scientists themselves use this method of carrying a portable atomic clock, which has been carefully synchronized with NBS 6, to Paris for comparison with the time scale kept by the International Time Bureau. Few persons know of the careful planning, exacting effort, and scientific know-how needed to keep clocks around the world synchronized within microseconds. In the next chapters we shall explore some of the needs for this degree of precision, and the wonders of modern technology that are possible because of it.

II

Putting Clocks & Time to Work

8

Order in the Air Waves

Who are the "fans" who tune in short-wave radio stations like WWV and WWVH? We've already mentioned a few, such as scientific laboratories, electric power companies, TV and radio stations, and navigators of ships and aircraft. There are many others. Jewelers who sell and repair clocks and watches, camera makers and repairers, persons who make and repair electronic organs and other musical instruments, and owners of private yachts and pleasure boats and small aircraft. Navigation schools teach courses in using radio broadcasts and other sources of precise time, and manufacturers make radio receivers designed to tune in just the frequencies used by WWV and WWVH.

Amateur or "ham" radio operators listen regularly to these stations. They wish to record precisely the times of their contacts with other operators, perhaps halfway around the world. More importantly, they must be sure to keep their own broadcast exactly "on frequency." As the diagram shows, the short-wave "band" is not very wide; and like other bands, it is extremely crowded. The penalty for straying outside one's assigned frequency could be loss of license to operate. So part of the ham's "housekeeping" duties include tuning in a station he or she knows is precisely on *its* assigned frequency, and checking equipment against it.

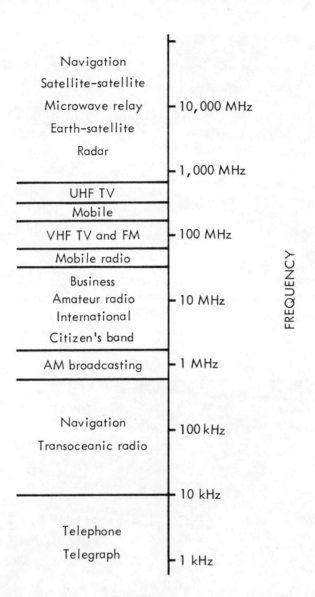

How does the signal travel almost instantaneously around the world? In 1901 Guglielmo Marconi, an early experimenter with wireless communication, detected three faint signals on his crude radio receiver. The receiver was located in the southeastern corner of Canada, and the signals had crossed the Atlantic Ocean from England. Scientists were

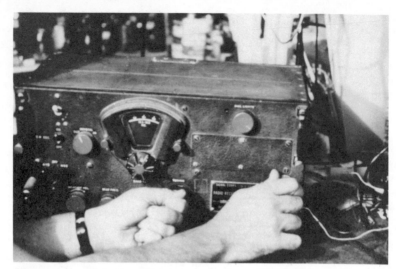

Ham radio operators listen regularly to NBS time and frequency stations.

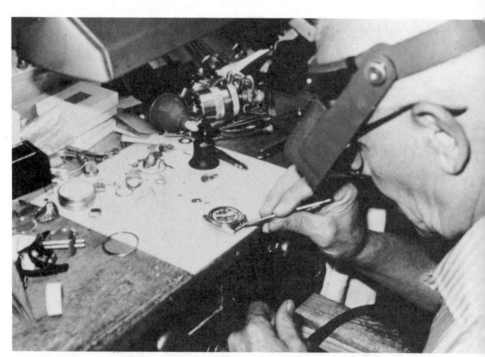

Watchmakers need precise time information.

Organ builders need precise frequency information.

Electronics equipment operation depends on both time and frequency information.

perplexed. Radio waves, like light waves, travel in straight lines, and do not follow the curvature of the earth. Marconi's experience was as startling as seeing a friend standing behind a tree.

There was another curious thing. Marconi received signals from greater distances at night than in the daytime. It seemed as if the radio signals were somehow affected by the earth's atmosphere. Some scientists, working independently, came up with the idea that high in the earth's atmosphere there was a mirror-like region that could reflect radio signals back to earth.

Over the next 25 years many experiments were conducted to find the nature of this atmospheric mirror. One man in particular, Sir Edward Appleton, gave so much to our understanding of this mirror—which we now call the *ionosphere*—that he received a Nobel prize for his work. Appleton's experiments showed clearly that the upper atmosphere consists of a number of reflecting layers, one on top of another. The layers are produced by the sun's rays shining on the almost vacuum-like atmosphere many kilometers above the surface of the earth. Because the strength of the sun's rays changes with time of day, season, sunspot activity, and other factors, the ionosphere changes its shape and moves up and down; so the "path delay" time of a radio signal bouncing along between the earth and the ionosphere is quite variable. This is why radio time signals from stations like WWV have limited accuracy.

It didn't take mankind long to realize the value of being able to communicate almost instantaneously over long distances without wires. Today any modern nation would be paralyzed if it lost its radio communication ability. Thousands of transmitters generate signals that are heard by millions of listeners in almost every part of the globe. But this could not have happened without careful planning and cooperation nationally and internationally—or without the tools provided by time and frequency technology.

Let's look more closely at how radios work. The *antenna* of any radio receiver is bathed constantly in a sea of radio signals, some weak and some strong, and at many different frequencies, coming from all over the world. What do these signals "look" like? Of course we can't actually see a radio signal, but we know from the way it acts what it must look like. Suppose we look first at a signal coming from a transmitter when nothing is being said—during a short interval at the time of a station break, perhaps. We would see a wave passing by at nearly the speed of light. It's a simple-looking wave, which we might picture as in Fig. 1. It fluctuates constantly in strength, or *amplitude,* so that it has a kind of pulse, or beat, and we can count these pulses. Suppose we count the waves passing by us in one second; maybe we count one million waves. Then we say the *frequency* of this radio wave is one million cycles per second—one million hertz, or one *megahertz* (MHz). This signal is silent—the frequency too high to hear.

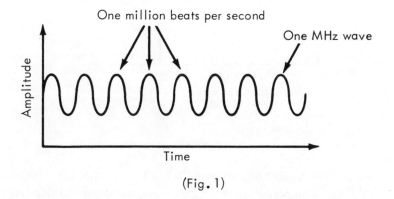

(Fig. 1)

After a moment of silence we hear a simple tone; perhaps it marks the beginning of a new hour. Let's suppose the frequency of this tone is 1,000 hertz—one kilohertz (kHz) —a frequency that the human ear can hear. When we

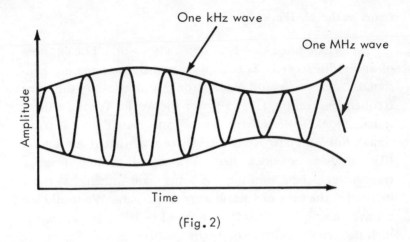

One kHz wave

One MHz wave

Amplitude

Time

(Fig. 2)

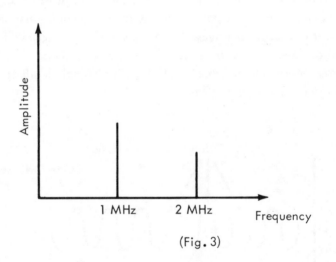

Amplitude

1 MHz   2 MHz

Frequency

(Fig. 3)

look at our wave now, we see that it has changed its shape. "Riding" along with the 1-megahertz signal is our 1-kilohertz tone, as shown in Fig. 2. It's as though we took our original 1-megahertz signal and passed it through a gate that swings back and forth a thousand times a second, not completely closing, but closing just enough on each swing to shape the signal as shown in this figure. This shaping process is called

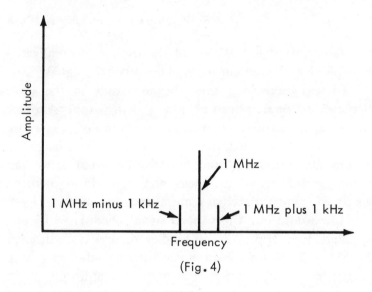

(Fig. 4)

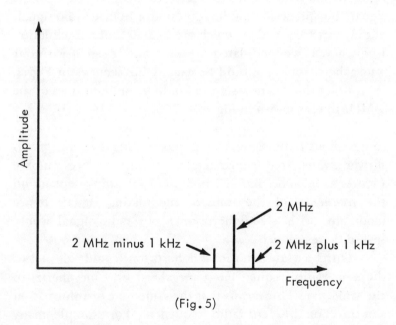

(Fig. 5)

modulating the signal. We could also use more complicated signals, such as the human voice, to modulate the signal.

These figures show how the amplitude of the wave fluctuates in *time,* as seen at our "viewing" spot. Another simple way to picture radio waves is shown in Fig. 3. Here we have a vertical line centered at 1 megahertz. The *height* of this line shows the amplitude of our radio signal. If we had another, *weaker* signal at 2 megahertz, it would be pictured by a shorter line centered at 2 megahertz, as shown in Fig. 3.

How would our 1-megahertz signal, modulated by our 1-kilohertz tone, appear in this kind of picture? It would look like Fig. 4. Here we see three vertical lines—the same long one at 1 MHz, and *two* new ones, with less amplitude, at one kilohertz more than and 1 kilohertz less than our 1-megahertz signal. The process of modulation produces new radio signals at different frequencies, which are related to the modulation frequency. If we modulated the 2-megahertz signal with our 1-megahertz tone, it would be exactly the same as in Fig. 4, except that the two new signals would be on both sides of the 2-MHz line, as shown in Fig. 5.

This simple example points out a number of important things about radio signals. First, modulating a signal *spreads* it over a wider frequency range, as we can see if we compare Figures 3 and 4. Second, the *amount* of spreading depends on the *frequency* of the tone or modulating signal. If we modulated with a 2-kilohertz tone, our radio signal would spread twice as much.

Third, we saw that the 1-kilohertz tone could ride on two different radio frequencies—1 megahertz or 2 megahertz—in the same way. This means that the same tone or information can travel on different radio frequencies. For example, many different radio stations may broadcast the same program at the same time—a speech by the president, perhaps. When this happens we can hear the same speech on several different stations, represented by different locations on our radio dial —or different frequencies.

The center radio frequency signal on which the information signals ride is called the *carrier* signal, and the signals surrounding it are called *side band* signals. The kind of modulation we've been discussing is called *amplitude modulation*—or what we know as AM radio—because we have been changing or modulating the amplitude of the signal.

But there are other kinds of modulation. In one kind, the *frequency* of the carrier signal swings back and forth in response to the modulation signal—or in our example, in response to the 1-kilohertz tone. This kind of modulation is called *frequency modulation,* the kind used in FM radio. The important point for us to remember is that no matter what kind of modulation is used, the carrier signal is spread out in frequency by the modulating signal.

Now that we see how information travels on radio waves, we can understand how radios work. Radio receivers simply "undo" what was "done" at the transmission station. Many incoming signals at many different frequencies are picked up by the radio antenna. When we "tune" a radio, we are simply telling the receiver which one of the many signals we wish to listen to. The receiver takes the selected signal and separates the information—that is, the side band signals—from the carrier signal. The receiver must also greatly amplify the signal before we can hear it, since the radio signals arriving at the antenna are very weak after their long journey.

With all this discussion of frequency, it's not surprising to find that frequency-generating devices, or *oscillators,* are important in broadcasting. An oscillator is similar to a resonator in a clock; it vibrates, or oscillates, to produce a specified frequency. A radio receiver must contain an oscillator so that the listener can "tune in" the frequency of the station he wishes to hear. And radio stations must also have oscillators so that they can keep their broadcast on their assigned frequency.

In the early days of radio, oscillators were not very good; the radio stations constantly drifted away from their assigned frequencies. Once in the 1920s the dirigible *Shenandoah* became lost off the east coast of the United States during a winter storm; all New York radio stations had to go off the air so that signals from the *Shenandoah* would not be covered up while the airship was guided to safety.

As the numbers of radio stations grew, it became more and more necessary for each one to stay on its proper frequency. If it did not, it might stray into another station's "territory" and cause interference. One of the earliest uses of standard frequency broadcasts, such as those provided by WWV, was to keep stations in their proper "lanes." Radio station engineers could check their oscillators against the standard frequency broadcasts and adjust their equipment when necessary.

But even when stations stay on their assigned frequency, there's another problem. Many radios tune in a wider range of frequencies than is really needed to get all the information from the side bands; the result may be the picking up of faint signals from another program from a different station broadcasting on a frequency near that of the tuned-in station.

One way to solve this interference problem might be to make the frequency separation between stations so wide that one is not likely to drift into another's territory. But this is somewhat like making automobile lanes one kilometer wide so that people won't drive off the road. It's very wasteful. Radio space, like water and air, is a limited natural resource; and like other limited resources, it must be kept as "pure" as possible and used efficiently.

In assigning frequencies to radio stations, the Federal Communications Commission leaves a gap called a *guard band* between stations serving the same area, to allow for the possibility of drift. One way to make room for more stations is to reduce the "size" of these guard bands. And the only way to reduce the size of the guard bands is to provide even better frequency control equipment at radio stations to cut down on the drift, and better frequency devices in radios so that they can be tuned to just the station one wishes to hear.

New commercial radio stations come on the air almost daily, and citizens band (CB) and mobile radio have grown explosively in the last few years. How can all of these demands be met? Someday most of the commercial TV and radio signals may be carried to our homes over cables, relieving some of the congestion in the airwaves. But of course radios and transmitters in moving cars, airplanes, and ships cannot be connected by wires, and perhaps new technology will have to be developed.

We have seen how better frequency control in both transmitters and receivers allows us to pack stations closer together. But techniques are also being developed to squeeze more information into the same "piece" of radio space. All of these depend heavily on time and frequency technology, but we shall mention just one here. When people speak, there are small gaps in their talk. By using very precise timing techniques, it is possible to sneak information into these tiny unfilled gaps. In Chapters 11 and 12 we shall learn more about how this is done, and also how time and frequency help to keep our conversations secure from unwanted listeners, whether we are talking by radio or by wire.

9

The Clock in Your Television Set

SUPPOSE YOU WANT to send a large photograph to a friend, but the only envelope you have is too narrow. You consider several solutions, and finally decide to cut the picture into a number of narrow strips that will just fit the envelope, as shown in the figure. To make things easy for your friend, you number the strips from top to bottom, so that he can easily put the picture together again.

This may seem like a strange procedure, but it's really very similar to the scheme used to send television pictures. The camera in the TV studio *scans* or divides the image in front of it into a large number of horizontal strips—525, in the United States system. As the camera scans the strip, it changes it into an *electrical signal.* The signal consists of a long train of horizontal strips, each strip about 60 micro-seconds long. The total train of 525 strips is only 1/30 second long.

After it has scanned the 525th strip, the scanning device in the TV camera returns to the top of the image and repeats the process. So the whole picture "starts over" 30 times every second. This is far too fast, of course, for the human eye to detect; it is faster than the change of frames in the 35-mm motion picture films shown in movie theaters, which run at

Strip # 2

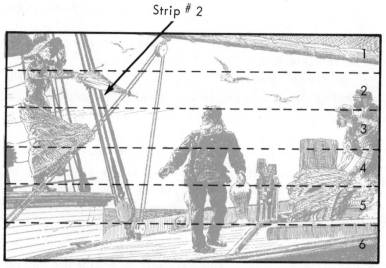

Photograph

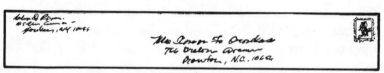

Envelope

the rate of 24 frames per second. But sending and receiving a televison image is far more complex than recording an event on film for later projection, and requires much better timing.

A TV receiver—the home set we're all familiar with—is kind of a TV camera in reverse; it takes the horizontal strips, one by one, and puts them back together to form a picture on the TV screen. It sounds very simple, but anyone who has peered inside a TV set with its mazes of wires, tubes, transistors, capacitors, and resistors realizes that carrying out the basic idea is not simple at all. Some circuits convert the electrical signals back into light images, or the picture that we see. Others change the audio part of the TV signal into the sound that we hear. One very important job of the TV set is to make sure that the horizontal strips of the picture are reassembled correctly. Clocks are a vital part of this job.

Before it sends each strip of the picture, the TV transmitter sends a timing or *synchronization* pulse—called a "sync pulse" by engineers and technicians. When this pulse arrives at the TV set, it tells the set that a strip of picture is coming; this keeps the TV set in step with the TV camera. There is also a special sync pulse that tells the TV set that the last—or 525th—strip has been sent and that the next strip should be inserted at the top of the screen. If the TV set gets "out of sync" with the camera, then the picture "rolls" and the horizontal strips run off in diagonal directions across the screen. The "horizontal" and "vertical" adjustment knobs on TV sets are there to correct these problems.

Quite often electrical "noise"—perhaps from a passing automobile—disturbs the sync pulses so that they are no longer able to keep the TV set in step with the camera. To make this problem as small as possible, manufacturers put clocks in TV sets; the clock keeps the TV set in step with the studio camera when the sync pulses are too "noisy" to do the job. No clock is perfect, of course; so the "good" sync pulses are used to keep the clock running at the correct rate. In this way the sync pulses and the clock work together to keep the TV set in step with the camera.

About 15 percent of the communication capacity of a TV signal is used just for timing. If TV sets had better clocks, they could get by with fewer sync pulses, because the clocks could run longer between "resettings." This fact shows us an important connection between clocks and the efficiency of timed or *synchronous* communication systems, which we shall be exploring later: The better the clocks in the system, the less the portion of the signal that must be used just to keep the sender and receiver in step, and the greater the portion of the signal that can be used for the message itself. In principle, TV sets with perfect clocks could get by without any sync pulses at all.

But the better the clock, the more it costs. So there has

to be some balance between the percent of communication signal used for timing and the cost of the clocks. In the future, however, we may expect electronics improvements and mass production of better clocks to lower the price per clock. As clocks become less expensive and competition for use of the radio spectrum grows, it may be less expensive and more effective to put better clocks into TV sets.

Sometimes the sync pulses in TV signals are used for timing or synchronizing clocks other than those in TV sets. To illustrate, let's suppose that you and a friend have "Super Special Space-Age" watches that tell time in microseconds. You wish to synchronize your watches very accurately, and both of you can receive the same TV channel in your city. Each of you also has a "black box" attached to your TV set that lets you see the sync pulses. Let's suppose that your house is 10 kilometers and your friend's house is 20 kilometers from the transmitter. You could find these distances from an accurate map.

TV signals travel at nearly the speed of light, which means that the signal takes about 3 microseconds to travel one kilometer. Since your house is 10 kilometers closer to the transmitter than your friend's house, a particular sync pulse will arrive at your house about 30 microseconds earlier than it will at your friend's house: 3 microseconds times 10 kilometers equals 30 microseconds.

If your two watches are synchronized, then your friend will receive the sync pulse 30 microseconds later than you receive it, as shown by your two watches. Suppose, though, that your friend's watch is ahead of yours by 5 microseconds. Then your two watches will show a 35-microsecond difference in the arrival time of the sync pulse. Since they should show just a 30-microsecond difference, you know that the watches differ by 5 microseconds.

Of course there are no wrist watches on which we can read time in microseconds, and there's no common way to observe the sync pulses. But scientific laboratories and other users can and do use this method to synchronize very precise clocks. They use an *oscilloscope,* which translates sound waves into light on a TV-like screen, to look at the pulses. Any "pulsed" radio signal, in fact, that can be received at two or more locations a known distance from the transmitter, can be used to synchronize clocks at those locations. TV sync pulses are better than most other signals for this purpose because they are very sharp and narrow, and so their travel time can be measured very precisely.

The TV sync pulse, by itself, does not give any *date* information, of course. It only allows us to determine the *difference* between clocks. The National Bureau of Standards, however, has experimented with another kind of TV time system that does give the date. In this system a special code is sent along with the normal TV signal from the station, which gives the year, month, day, hour, minute, and second. This TV time signal is "hidden," so that it does not interfere with normal TV signals; in fact, only TV sets with an extra circuit can detect and display the time.

The system is really very neat. We have said that the TV picture consists of 525 strips, or "lines," as they are usually called. Actually a few of the lines are not used to send the picture; so they can be used to send other information. These unused lines are normally hidden under the frame or mask that surrounds the TV screen; but when your TV picture rolls, you can see these unused lines as a horizontal black bar that rolls with the picture. The bar is black because no information is contained in any of the lines that make it. The TV time signal is sent in one of these unused lines in the black bar, and then is decoded and displayed in the corner of the screen by a special electronic circuit or "chip" in the set.

The equipment developed to send the time signals can also be adapted to send other kinds of information, such as *captions.* The captions appear on the TV screen in much the same way that English captions are used along the bottom of the picture in some foreign films. Again, only TV sets equipped with the special circuit can pick up and display the captions. The system was adopted by major networks in 1979

Captions broadcast by the network master station appear only on TV sets equipped with a special device that decodes the signal.

to provide captioned TV programs for deaf persons; one of its main uses is to help deaf children learn to read. Its big advantage is that the large majority of viewers, who do not like to be distracted by captions appearing in the picture, don't have to have them.

The National Bureau of Standards' experiments were first intended to explore the use of TV for time broadcasts, but the resulting captioning system is a good example of something that often happens in science and technology. A project directed toward one goal often leads to important results in an unexpected area. Whatever the goal, and whatever the eventual use of the system, television as we know it today and as it may develop in the future would be impossible without very fine clocks to keep the strips of picture moving in an orderly way. And like radio stations, TV stations also depend on precise frequency standards to keep their broadcast on its assigned channel.

10

Keeping the Electric Companies in Step

ONE OF THE MOST dependable time sources we have is the common electric wall or desk clock. Fancy and plain, large and small, there's a style to fit every taste, with a choice of dial face or digits. These clocks are so inexpensive that most homes have several. They keep such good time that, except for a power failure within the user's home or throughout an area, they could display time accurately within a few seconds for years without ever being reset. Except, of course, for the twice-a-year changes to accommodate daylight saving time.

A little thought about what we now know about clocks, however, will show us that this familiar time piece is not really a clock at all. It's just a device for counting the *hertz*, or cycles of "alternating current," generated by the power company, and displaying the count. This so-called clock has no *resonator* of its own, and no energy source. The clock is useless until it is plugged into the electric system. Its mechanism is designed to "run" at the "line frequency" maintained by the power company, which throughout the United States and Canada is 60 hertz. In Europe and most of the rest of the world, the standard line frequency is 50 hertz, and clocks for these areas must be designed and manufactured to run at that frequency.

This neat time piece, then, is simply a "slave" to the "master" clock maintained by the power company. If the master clock were for some reason to start running inaccurately, all electric clocks plugged in anywhere in the system would show this same inaccuracy. In countries where power companies are less carefully operated than they are in industrialized nations, electric clocks keep time very poorly.

In some large factories, business complexes, and other buildings where there are many clocks, there are even slaves to the slave clock. The wall clock in your school room may be a slave to a master clock in the principal's office; and this, in turn, is a slave to the master clock at the power generating plant. Such systems, like many other electric and electronic systems, are very expensive to install; but they are time savers, and therefore money savers in the long run. Whenever there's a change, to daylight saving time, for instance, the building custodian can change just the master clock, instead of going around to every room to change each clock. Another advantage is that all clocks in the system automatically show exactly the same time.

How does the power company maintain its fine master clock that we all rely on and take for granted? It must, of course, have a source of time and frequency that *it* can refer to; and this source must be even better than the time information the power company supplies. So it has its own quartz-crystal frequency standard. One of the duties of a technician at the power plant is to check this instrument routinely, comparing it with an "official" time and frequency source, such as the radio broadcasts from one of the National Bureau of Standards stations. He keeps a careful record of the performance of the company's frequency standard, and resets it when that is necessary. This frequency standard controls the generators, so that they put out electricity at a very accurate 60 hertz.

Perfect *stability*, however, is impossible to maintain. An accident or equipment failure in one part of the system may put a momentary heavy demand on the rest of the system. Or many persons may switch on their TV at about the same time to look at a special broadcast. This sudden drain of electric current may lower the line frequency briefly, until the generators increase the output to meet the demand. Frequency may vary by a few tenths of one hertz. The situation is much like turning on all the water faucets in a house at once, thus making a larger demand than the system is able to supply; so less water runs out of each faucet, and the water runs more slowly.

Electric clocks, of course, run "slow" when the frequency drops, and accumulate the error until the normal frequency is restored. The error would then remain, except that the power company carefully "erases" it. The technician keeps precise records of just what the error is, and then increases the frequency slightly above the normal 60 hertz until the error is removed. For instance, if the frequency drops to 59.9 hertz for several seconds, he would increase frequency to 60.1 hertz for the same length of time. The time error is rarely more than 2 seconds.

Keeping electric clocks telling the right time, however, is only a very small part of the role frequency plays in electric power generation and distribution. It is really just a sort of side effect, in fact, or "fallout benefit" that comes from the company's own need for accurate frequency measurement.

As the batteries in your cassette tape player become weak, the speed of the tape slows down and the music "flats." If the speed were too fast, the pitch of the music would be higher than normal, or "sharp." This distortion is due to a "Doppler-like effect," or the lengthening or shortening of sound waves. You've probably studied Doppler effect in your science classes, and we shall be discussing it more fully later. When you discard the weak batteries and plug your player into an adaptor, and then into the household electric system, the tape runs at the proper rate and the pitch of the music is right.

Millions of motors in everything from tape and record players and electric shavers and toothbrushes to electric typewriters, washing machines, and heavy machinery in factories are designed to run on electricity supplied at 60 hertz. If a family moves from New York or Shreveport to Seattle or Los Angeles, they can expect their vacuum cleaner, hair drier, and food processor to work just the way they did back home.

This was not always true. Up until World War II there were a few places in the United States—even in heavily populated areas—where electricity was supplied at 50 hertz. But having to discard appliances, including electric clocks, or have the motor replaced with one built for the different frequency of the local system was an expensive nuisance. A motor designed to run at 60 hertz becomes overheated if it is plugged into a 50-hertz system, because it "tries" so hard to work up to its normal speed that it not only operates inefficiently but wears out sooner than it should. Now all appliances can be used properly in every part of the United States and Canada.

Frequency can easily be measured in any part of a power system. So monitoring the frequency at various points is a good way to "feel the pulse" of the system and check its health. A drop in line frequency is a distress call to the generators for more energy.

An important characteristic of power generation is that the alternating-current electricity supplied by power companies cannot be made ahead of time and stored like coal in a pile or like water in a reservoir—or even like the small amounts of electricity stored in batteries for "direct-current" on-the-spot use. Alternating-current electricity must be generated as it is used. So to provide dependable service, many power companies have formed regional "pools." Some networks extend over very large areas covering many states. When demand is very heavy in one area, the power plant there can draw some electricity from a neighboring area where demand is less at that time. Demand is usually heavy, for instance, when many families arrive home from work and are cooking their dinners, running their air conditioners, and watching TV. In the time zone just to the west, however, it's an hour earlier, and these activities have not yet begun. So power companies in this time zone can help meet the peak demands of their neighbors.

Or there may be a severe storm in a certain area, not only causing heavy demand for electricity as people turn up their heaters and use extra light, but maybe causing trees to fall across power lines or knocking out service with flashes of lightning that destroy or disable electric power transformers and other equipment. Again, pools of electrical energy can be tapped from surrounding areas to aid the stricken regions until power is restored.

If an over-demand for current in your home blows a fuse or "kicks" the circuit breaker and leaves the house without electricity, the first step in restoring service is to find out where the trouble is and what caused it. There may be a "short circuit," caused perhaps by a frayed electric cord; or maybe there were too many appliances operating at once on the same circuit.

Finding the cause and location of the breakdown is also the first step the electric company takes to restore service. Often the sequence of events that leads to a power failure happens so fast that no one can be sure what was cause and what was effect. To sort out the events and show the order in which they occurred, very precisely synchronized "event recorders" are scattered throughout the network. Most event recorders must be synchronized within one millisecond or better, and for very detailed analysis, synchronization must be within a few microseconds.

Sometimes locating the place where the failure occurred is easy—when a truck accidentally breaks off a power pole, for instance. But sometimes the cause is hidden. A main line may be down on some remote mountainside, or something may have happened to an underground cable or a transformer in a manhole somewhere. Accurate time throughout the system helps to locate the problem area. When such an event occurs, a surge of electric current, or *transient,* spreads over the power system. By timing the arrival of this transient at various points throughout the system it is possible to pinpoint the location of the trouble. As an example, a tree may fall on a power line connecting two power stations. The transient first appears at the event recorders at the power stations nearest to the tree, and then shortly after at the other stations. The difference in arrival time of the transient can be used to find the location of the trouble.

In the last few years you've probably heard about—or perhaps even experienced—very widespread "blackouts," power failures that affect whole areas including several states. What has happened? Often the problem starts with a sudden heavy demand, larger than local generators can meet, in one small area. So this area calls on the pool for help. But if nearby units in the pool have little to spare—or if two or three units call for help at the same time and the power lines aren't big enough to carry the extra load—then the demand becomes more than even the pool can deliver.

Getting the system running again is an exacting and sometimes slow process. If the engineers just turned everything on again at once—and all the people's refrigerators and air conditioners and TV sets instantly demanded electricity—the system could not meet the demand and the blackout would happen all over again. *Starting* motors and other equipment takes lots more electricity than keeping them running after they're started.

Power must be restored gradually, and both clocks and frequency standards are essential to this orderly process. As each idle generator is started up, it must be running at the same frequency as others already "on line" before it can be connected into the system. If it is running too slowly, current will flow into its windings from the rest of the system to try to bring it up to speed; if it is running too fast, extra current will flow out of its windings to try to slow it down. In either case, these currents may damage the machinery.

Besides running at the same frequency as the rest of the system, the new generator must also be in step, or in *phase,* with the rest of the system. We can understand phase by thinking of a marching band. If the marchers all step in time to the drummer's beat, they are marching at the same rate, or *frequency.* But if some left feet and some right feet move forward together, the marchers are "out of phase." If a generator coming on line is out of phase, strong currents try

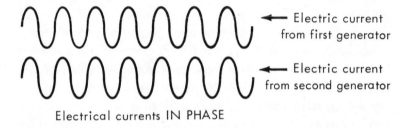

Electric current from first generator

Electric current from second generator

Electrical currents IN PHASE

to bring it into phase, again possibly damaging the machinery. Electric companies have developed devices that allow them to make sure that a generator is connected to the system only after it is running with the correct frequency and is in phase. When there has been a massive blackout in an area, the generators are brought back into service one at a time, and customers must usually limit their use of electricity until the whole system is operating normally again.

As the cost and demand for electrical energy increases, it is more and more important to distribute it efficiently and reliably. Time and frequency technology is a necessary part of reaching this goal.

11

The Secrets of
Secret Messages

THE TELEPHONE is one of the most useful devices men have invented for communicating with each other. But persons on the same party line can listen to what others are saying, and even private lines can easily be tapped. Radio is much less private, as anyone who has used a CB radio or even a walkie-talkie knows. Whoever has a radio receiver can tune in on the conversation. So how can we send private messages?

Most children have made up some kind of secret code to keep their message secure from prying classmates and inquisitive teachers. One simple idea is to substitute each letter of the alphabet with some other letter. Such a code works quite well for a short message that is not important to many people. But anyone with a little experience in breaking codes can easily decipher it. The coding methods we shall discuss here depend on scrambling and distorting messages electronically so that they may pass by spying eyes and ears unnoticed. Or if they are noticed, they will have no meaning except to those who know the secret of unscrambling them.

Clocks play an important part in communication systems where secrecy is necessary, and it's not just spies or police checking suspected criminals who need privacy. Every day, stock market tips, financial statements, medical records, contract offers to sports and entertainment stars, and many other kinds of information must be delivered privately to various persons who need to receive them quickly and accurately.

But even when privacy is not important, clocks and frequency standards provide the basis for today's communication systems. So let's put codes and secrecy aside briefly while we see how ordinary messages travel electronically.

When we speak into a microphone or telephone, we produce a continuous electric signal; it might look something like the signal shown in the sketch. As we vary the pitch and loudness of our voice, the signal frequency and strength also vary. We could record this signal on magnetic tape and play the tape back later to reproduce the original speech. Or we could send the tape to a friend and he could play it on his own tape recorder at the same speed we recorded it, to get our message.

Messages recorded in this *continuous* way are called "analog" signals. Most phone conversations travel this way. But because we now have electronics and very fine clocks and frequency standards, there is also a *discontinuous* way to record messages, and the result is called a "digital" recording. Usually the digital recording is made by "sampling" the analog signal many times per second, as shown in the sketch. Each sample is given a number that represents the strength of the signal at that instant. In our sketch we have shown a piece of the signal that varies in strength from 42 to 112. We could now record these numbers on tape to represent the original speech pattern. Then by changing the numbers back into what each one represents, we could reconstruct the message.

The usual process, however, involves one more step, to change the numbers into what is called a "binary code." This process translates the speech pattern into a long string of o's and 1's, or "bits." One way of doing this is really very simple. Suppose that all of the numbers used to represent our voice

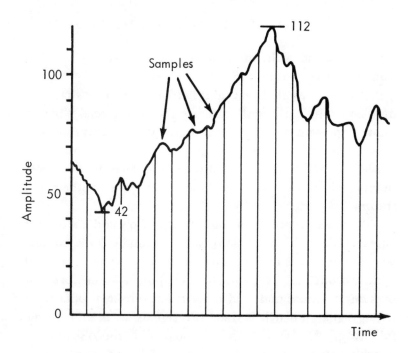

Samples

Amplitude

112

100

50

42

0

Time

message vary in strength between 0 and 128, and that we want to change our first sample, with a strength of 71, to just a string of 0's and 1's. We begin by dividing all of our numbers—0 to 128—into two equal groups; all numbers 0 through 64 are in one group which we'll call 0, and those from 65 through 128 are in the other, which we'll call 1. We then ask, "Is 71 in the first half or the second half?" Since 71 is greater than 64, it is in the second group, and we note this fact by writing down the number "1."

Now we divide the group of numbers 65 through 128 into two equal groups. And again we ask the question, "Is 71 in the first half or the second half of this new division?" Since it is in the first half, we write down the number "0." We continue in this way, dividing the remaining group in half and writing down a 1 or a 0 depending on which half of each group our number 71 is in. When we have written seven numbers, or digits, we find that we have 1001001, a code symbol that shows the number 71 in a way that fits no other number between 0 and 128.

Although we shall not do it here, we can prove that each number from 0 through 128 has its own special "binary digital representation" of seven digits. We call this representation "binary" because all numbers consist of just two symbols, 0 and 1. You have probably learned about simple binary arithmetic, with its base 2, in your mathematics studies.

Next we record all of our strings of 0's and 1's on our tape recorder. There are many ways this can be done, but one very simple way is to have some device that puts out a signal with two unequal strengths. If a 0 is put into this device, it puts out a weak signal; and if a 1 is put into it, it puts out a strong signal. Then it is these two signals, weak and strong, following each other exactly in step with the 0's and 1's, which we record on tape. In a way, we're back to our familiar string of beads, with the 0 and 1 beads speeding through the *now* gate, one after another.

We have now changed our original *continuous analog* message—our voice message—into a *discontinuous, two-level digital* message. It's hard to imagine that all this could be happening to everything we say as we visit on the telephone with our friend who may be just around the corner, but that's how advanced systems work today.

Why have we gone to so much trouble? One reason is that computers can easily store and work with 0's and 1's; we'll discuss this in more detail in Chapter 14. But there is another reason that's important from a communication standpoint. Binary signals representing many different telephone conversations can be sent over a single wire or communication channel at the same time. The figure shows how this is done for just three telephone conversations. But in actual practice, *hundreds* of conversations can go on over the same wire—or more recently, tiny glass fiber—at the same time.

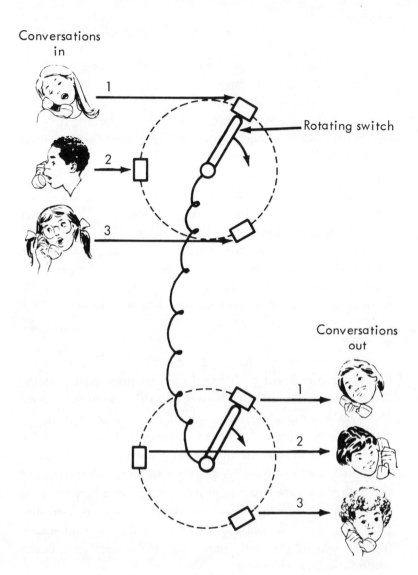

Conversations in

1

Rotating switch

2

3

Conversations out

1

2

3

In our simple diagram the three signals are fed to a rotating sampler-switch, which samples each conversation in turn and sends the digital samples down the same wire to another rotating switch that accepts the digits as they come in, separates them, and sends them along three separate lines, each to the intended receiver. Here the digits may be electronically converted back into the sound we recognize as someone's voice. Or in cases where it's simply one computer "talking" to another computer, there is no need to translate the "computer talk" into sound at all.

Now we see why *clocks* are so important to this communication system, which is called a "synchronous communication system." *Date,* of course, is of no importance to the system. But the two clocks at the switches must be very precisely *synchronized.* If the rotating switches are not running at exactly the same rate and in the proper alignment, the information will be sent to the wrong receivers. Or, worse, the three conversations won't be separated correctly, and we will have changed them into gibberish. In some very fast systems, the information may be flowing at a rate of several million bits per second. So the clocks must stay together within a fraction of a millionth of a second.

The system we have just described is called "time division multiplexing." It has a kind of sister system called "frequency division multiplexing." In this system the several messages flowing along the same line are kept separate by sending each one at a different *frequency,* as shown in the figure. If we are to separate the messages correctly at the receiving point, we must have frequency standards at both ends of the path which are running at exactly the same frequency, to make sure that the signals are not distorted or converted into gibberish. Time division and frequency division multiplexing are two of the most important techniques in modern communication systems.

Now that we understand a little bit about how messages travel electronically, let's return to our tape recorder and record some secret message. Then to send our message, we'll run our tape recorder very fast, so that the message, which may be several minutes long, is reduced to a "microdot" of sound—perhaps less than one second long. Such a microdot might pass unnoticed if it were slipped into an ordinary radio broadcast. But then we must spread the sound out again at the receiving end if we are to understand the message. You've probably already guessed that we need a frequency standard so that we can slow down the sound by just the right amount. In addition, we might know in advance that the microdot of sound is to be sent at a particular instant in time; so now we also need a clock, synchronized with the clock where the message originates, to tell us at just what time to listen.

The sound microdot suggests another possibility—a sound "*macro*dot." Now we play the message at a very *slow* rate, so that all we hear is a low grumble that might be mistaken simply for "noise" on the telephone or radio broadcast. Again we need a frequency standard connected to our receiver to speed up the message by just the right amount so that we can understand it.

We can think of even trickier ways to scramble and hide sounds using timing technology. Suppose we speed up and slow down our tape recorder in a random way. Now to unscramble the sound we need to know the *time sequence* in addition to the *amount* of speeding up and slowing down. So of course, once again, we need a clock to do the actual unscrambling.

To make it even harder for an outsider to intercept our message, suppose we take our tape and chop it up into a great many different lengths and glue them back together in some known but random order. And then to make things really hard for the spies, let's transmit this mutilated tape while we also change the speed of the tape recorder at random. This shuffling of the tape into random short segments is called "time division scrambling," or TDS; it was invented in the 1920s to keep ham radio operators from listening in on private conversations and transmission of secret information.

We can look at these techniques in a little more general way by thinking about a radio station that can broadcast on perhaps five different frequencies at once. Suppose this station is on the air between 10:00 and 11:00 A.M., as shown in the figure. The X shows a particular message at 3

	10:00 AM		10:30 AM			11:00 AM
	m 1	m 2	m 3	m 4	m 5	m 6 →
1 MHz						
2 MHz						
3 MHz			**X**			
4 MHz						
5 MHz						

	10:00 AM			10:30 AM		11:00 AM
1 MHz				m 3		
2 MHz		m 2				
3 MHz	m 5				m 1	
4 MHz			m 4			
5 MHz						m 6

megahertz broadcast between 10:20 and 10:30, and the arrow shows a message one hour long broadcast at 1 megahertz. The figure shows this hour message broken up into 6 10-minute segments, M_1 through M_6. Now we could alter the *time* sequence of these 6 segments, as shown in the figure; and further, we could move them around in *frequency*, as shown. This is called "frequency hopping," and the term "spread spectrum" describes this kind of approach because the original message is spread out in frequency.

In modern spread-spectrum systems, the "cells" of information are broken up and spread in time and frequency under the control of a computer, so that often the message cannot be distinguished from noise. Thus an uninvited listener may not even realize that a message is being sent. The spread-spectrum system is one of the most advanced kinds of communication system in use today. Like other systems we've described here, it would be impossible without modern developments in time and frequency technology.

12

The Line Is Busy

IN 1876, Alexander Graham Bell spoke the first words into the first telephone: "Mr. Watson, come here; I want you." Mr. Watson was only one room away, in their Boston laboratory. Exciting as the invention was, few persons thought the telephone very useful, and many saw it as just an entertaining gadget. Bell and his friends had a hard time getting people to invest money in developing a telephone system.

Today there are over 120 million telephones in the United States, and about twice that many in the rest of the world. Almost any one of these phones can be connected to any other within minutes—often within seconds. How can each call made on these millions of instruments find its way—often through many connections and over thousands of miles of land and water—to reach the one receiver to whom it is directed? In the previous chapter we saw how a single message is coded and decoded on its journey. Now let's look at the vital role that time and frequency play in this remarkable system that is the backbone of today's electronic communication.

The most familiar part of the system, of course, is the telephone itself. But this is the smallest part. Spread out all

A telephone company microwave antenna synchronizes communication.

over the world—on telephone poles and microwave "repeaters," in pipes underground, in large and small buildings in villages and great cities—is a complex made up of the most sophisticated electronic equipment that man knows how to build. Operating along side and hooked into this equipment are millions of antiquated mechanical switches and other devices, some of them nearly as old as the telephone system itself.

When we lift the phone from its cradle, we signal the system that we are about to place a call. Dialing or punching out a number is simply giving instructions to the switching equipment in the network to set up a series of connections that will join our phone with the phone of the person we wish to call.

When we speak into the mouth piece, the *sound waves* generated by our voice are changed into *electrical signals* that travel down a short length of line called a "subscriber loop" to a nearby central exchange. The central exchange contains switching equipment that can connect our line to any other subscriber loop connected to the same central exchange.

But what if we wish to call a friend in another city, whose phone is not connected to *our* central exchange? In this case, the central exchange switch connects our call to a "trunk" line, which in turn connects our central exchange to other central exchanges. The diagram shows how a number of central exchanges are connected by a single trunk line.

A subscriber loop can handle only one conversation at a time, but a trunk line carries many calls at the same time. In the previous chapter we saw how Time Division Multiplex-

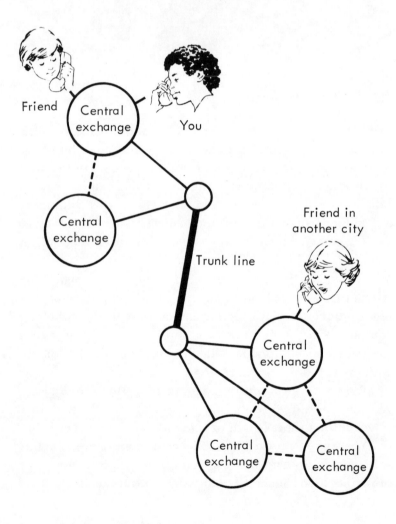

Friend

Central exchange

You

Central exchange

Friend in another city

Trunk line

Central exchange

Central exchange

Central exchange

ing (TDM) or Frequency Division Multiplexing (FDM) makes this possible. You will recall that TDM means that the continuous, or analog, electrical signal produced by the caller's voice is sampled to make a series of pulses which are then changed into a digital signal, a series of o's and 1's, or bits. These are then "interleaved" with bits from other phone conversations to produce a long line of bits representing many different conversations.

To provide a signal that can be recognized as the caller's voice, it is necessary to sample the analog signal 8000 times a second. Each sample is represented by 8 bits that occupy a "time slot" one microsecond long. One 8-bit time slot is called a "frame." A simple calculation shows us that sampling 8000 times a second means that there is a 125-microsecond time gap between every two samples. So the digital signal that represents a particular part of a message from one telephone consists of a series of 8-bit frames one microsecond long, with a 125-microsecond gap between frames. It is in this gap that many other calls can be interleaved or "sandwiched."

When FDM is used to send many different conversations down the same trunk line, each conversation is given a frequency "space" about 4 kilohertz wide, as shown in the figure. As we know from the previous chapter, both TDM and FDM require synchronized clocks at both ends of the path to separate the messages. If the clocks are not synchronized in frequency, then a frame of information may be lost and a "slip" will result. For voice transmissions there can be no more than one slip in 5 hours. Similarly, for FDM, a very small frequency offset between the clocks at the two ends of the path will result in distortion or loss of the message.

Keeping clocks in step when they are scattered from one end of the United States to the other is not easy. The Bell System's solution to the problem is a "master-slave" technique. A group of three master clocks located underground in a special room at Hillsboro, Missouri, controls

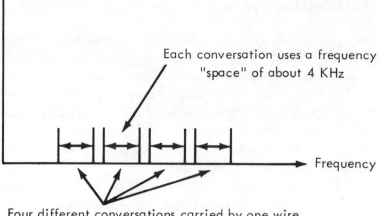

Each conversation uses a frequency "space" of about 4 KHz

Frequency

Four different conversations carried by one wire

many other clocks in the system. Very high-quality transmission lines carry the signal, almost without distortion, from the master clocks to the slave clocks, forcing them to stay in step.

As you have probably suspected, the master clocks are atomic cesium-beam standards. They are accurate to one part in 100 billion or better, which means that after 100 billion cycles, the signal is not in error by more than one cycle.

Why are there *three* master clocks? Well, if there were only one we would have no way to be sure that it was operating correctly. And with two clocks, if they were not keeping the same time we'd know that one was in error, but we couldn't know which one. So we need three clocks to detect a faulty one.

Finding ways to send many calls down the same line at the same time without getting them mixed up was a big achievement. But of course that's only one part of the process needed to put a call through from one telephone to any one of a hundred million others. If we are to reach one particular phone several thousand miles away, in a matter of seconds, there must be a very fast and accurate way to make the many connections needed between parts of the system. The two main kinds of *switching* schemes used by telephone companies are *Space Division Switching* (SDS) and *Time Division Switching* (TDS).

In Space Division Switching, a series of connections is set up through the telephone system so that the caller's phone is connected directly and continuously to the receiver's phone. It is as though there were a pair of wires running between the two phones. But with Time Division Switching, which is a much more recent development, the two phones are connected *intermittently,* one microsecond out of every 125 microseconds. This is just the right amount of time for digital signals, since as we have just seen, digital signals from a single telephone consist of a frame one microsecond long every 125 microseconds. So switches using the time-division method are ideal for handling digital signals.

But of course not all signals are digital. In fact, most today are analog. So when an analog signal approaches a time-division switch, it is first changed to digital form and then changed back to analog when it leaves the switch.

Why go to all this trouble of changing back and forth between analog and digital signals? There are a number of reasons, but two are most important. First, the latest version of these switches is all electronic, which means that it is fast and reliable—so fast that one switch can handle over 500,000 calls an hour connecting with more than 100,000 trunk lines. Second, because the information is in digital form, the switches can be controlled by a computer. Computer-controlled electronic switches can easily be altered by changing the set of instructions—or *program*—that tells the computer what to do. Two of these programable features, for instance, are having calls automatically forwarded to a friend's house if we are going to be away from home, and allowing the telephone company to do automatic billing.

Any device that is handling over 500,000 calls an hour must perform many operations with split-second precision at a very rapid rate. We might think of the electronic switch as a many-armed monster that snatches a packet of information one microsecond long—a frame—from an incoming line in one of its hands, and inserts it at the proper moment into the correct outgoing line. The hand must do this at the rate of one frame every 125 microseconds for the duration of the call. And if the monster has a hundred arms, it can handle 100 calls at a time.

It's easy to see that this many-armed monster would become a tangled mess in seconds if its movements were not coordinated and timed with the greatest precision. Its "clock" system produces a chain of pulses at the rate of 16,384,000 per second. And it must be perfectly in step with hundreds of other electronic switches spread over a wide area. If they are not in step, one switch will be trying to hand off information before another is ready for it.

But even when the switches are in step, there may be problems. Suppose the time it takes a packet of information to move from one switch to another changes; perhaps it takes a shorter time than normal. Then the packet will arrive ahead of schedule and be lost. To take care of this possibility, special devices called "buffer stores" have been developed to store a packet of information until the switch is ready to accept it. The buffer stores reduce the need for accurate timing throughout the system, particularly that part of the timing related to "travel" time from one switch to another.

The switching system "clock" consists of four quartz-crystal resonators in temperature-controlled ovens. The resonators are connected in such a way that any one of the four can act as a master to the other three. The circuit is designed so that if the master clock becomes faulty, its duties can be taken on by any one of the three remaining resonators.

Why not use the *best* clocks available—cesium-beam standards? One reason is that there are many switches throughout the system, and it would be very expensive to have a cesium-beam standard at every switch. However, the system is designed so that the quartz-crystal clocks can be slaved to better clocks, such as the cesium-beam system described earlier in this chapter.

One of the goals of the world's telephone companies is to build an international system in which any subscriber can dial directly any other subscriber in the world. And perhaps at a later time they will be able to provide worldwide picture-phone service. In the meantime, "the line" is very much busier than most of us realize, and none of the present services or future goals would be possible without today's modern clocks and frequency standards.

13

Smart Machines

RIGHT BEFORE our eyes—and yet almost unnoticed by most people—the world is experiencing one of the most dramatic changes in its machinery since the industrial revolution a century ago. We might call this change the "silent revolution." Complex machines with their wheels, springs, cogs, lever arms, belts and pulleys are giving way to silent electronic devices whose insides resemble stacks of aerial photographs—which, in a way, is what they are. Complicated circuits are first laid out with the aid of computers to produce an "image" of the circuit. Then photographic techniques reduce the image—or stack of images—to a "chip" that can fit inside a wrist watch or pocket calculator or any of hundreds of other machines in a space no larger than a postage stamp.

The tiny electronic newcomer is more reliable, longer lived, less power hungry, and generally far less expensive to produce than its mechanical cousin. Washing machines, microwave ovens, cameras, automobile ignitions, electronic games, and countless other devices that we all know and see every day march to the beat of the electronic clock, under the command of the microcomputer. So do organs in homes and churches, medical monitoring machines in hospitals, business

A single "chip" about 1 centimeter wide contains thousands of transistors.

machines in offices, and assembly-line machines in factories. We shall discuss computers themselves more fully in the next chapter.

It has been said that early-day mechanical clocks demonstrated and utilized almost every principle that later-day machines needed. In a similar sense, the electronic watch, with its motionless "movement" that has replaced the wheels, springs, and "ticks" of the mechanical watch, has led the parade of a bewildering array of new electronic devices. The popular pocket calculator probably deserves equal recognition.

The forerunner of the electronic watch was the electric watch, which was partly mechanical and partly electrical. On the outside this watch looked much like any other watch, but if you put it to your ear you found that the reliable tick had been replaced by the hum of a tiny tuning fork. The tuning fork, powered by a small battery, vibrated several hundred times per second and served as the resonator for the watch. The tuning fork—still a mechanical device—and the battery were two new ideas in watch construction, and the rapid vibration of the resonator—compared to that of previous all-mechanical watches—resulted in a watch that kept time accurately within about one minute a month.

The most recent development is the all electronic watch with no moving parts. The resonator is a tiny quartz crystal that vibrates many thousands of times per second; this gives us a watch that keeps time accurately within a few seconds a month. The time is displayed as digits, or numbers, on the watch face. Since the introduction of these watches, several features have been added, including display of the day of the week and the month. Some models have a stop watch provision, and some have an electronic alarm. Some watches now even contain a tiny calculator operated from a tiny keyboard.

The changes in watches have been so rapid and revolutionary in the last few years that it's hard to know what to expect next. There seems to be no end to man's ability to miniaturize electronic circuitry. Perhaps some future watch will contain a tiny computer, a two-way radio, voice recorder, and a miniature TV screen! By that time it will be stretching the imagination even to call this device a watch. Perhaps you can think of a better name.

Another familiar field in which electronics is making such rapid changes that new models appear almost daily is photography. Right after World War II there was a flood of marvelous new mechanical cameras from American, German, and Japanese factories. These cameras had features that made even the best of prewar cameras seem like museum pieces. But now, in the last few years these same manufacturers and others have put electronic devices into their cameras that make the fine mechanical cameras of the 1950s practically obsolete.

With the mechanical cameras of 30 years ago, the photographer had to set the shutter speed, lens opening, and focus manually—each a separate operation. Often he read a light meter near the subject he wished to photograph, to determine his settings. By the time the adjustments were made and the camera pointed toward the subject, the baby had probably already made a mess of the birthday cake and started crying. Or the subject—if it was a bird or beast—had vanished. Or a cloud had covered the sun. The right moment for the picture had been lost.

For some time cameras have been available that set either the shutter speed or the lens opening electronically, once one or the other had been selected. And there is now a camera that adjusts the focus. The proper focus for a camera depends, of course, on the distance between the camera and the object being photographed. With older cameras, the photographer had to judge the distance and set the focus accordingly. With slightly newer models, one could simply look through the view-finder and adjust the focus until the scene became sharp and clear. In the new camera, the focus is adjusted automatically.

This means that the camera must somehow "measure" the distance to the scene being pictured. The system for making this measurement is based on "echo distance measuring," which we shall discuss more thoroughly in later chapters. Briefly, the camera has a mechanism that emits several short bursts of sound, less than one millisecond long and at frequencies too high for the human ear to hear. These sound pulses travel toward the object being photographed and echo back to the camera, which electronically determines how long it takes the sound pulses to travel back and forth between the camera and the object.

The camera contains a miniature clock that times the travel time and the pulses. The greater the distance, the greater the travel time. After the camera determines the distance, a small motor in the camera adjusts the focus for

One of Rube Goldberg's elaborate machines.

this distance. The whole process takes just a few milliseconds.

Now that we've had a close look at watches and cameras, let's back away for a broader view of machines in general, and some of their common features. In the operation of any machine, a sequence of events must be carried out in an orderly way. In some machines, the completion of one event triggers the beginning of the next. Pressing the button for an elevator starts the machinery that brings the elevator to the floor where one is waiting; its arrival opens the door and holds it open for a certain number of seconds.

Cartoonist Rube Goldberg carried this basic principle to absurd lengths in his many cartoons that showed ridiculous and inefficient "machines" for accomplishing some task of little or no importance—such as locating your rubbers on a rainy day.

Fortunately there's another, and better, way to regulate machines. A good example is the timer on an electric dishwasher. When the operator turns on the machine, the timer takes over. As it clicks along, it carries out events of various time duration, one after the other. First perhaps a

heating element is turned on and water flows into the tub for 2 minutes. Then the faucet shuts off while the water spray-rinses the dishes for 3 minutes. Then the pump removes the rinse water for 2 minutes. After this process is repeated, the detergent cup tips the cleaner into the tub, the tub fills for 2 minutes, and the washing process goes on for 6 minutes. This water is pumped away for 2 minutes. Then two more rinse cycles and a dry cycle follow.

In both the Goldberg and the "clock-ordered" machines, the sequence of events is started by some event such as the falling rain or setting the timer on the dishwasher. Once the sequence is started, the machine "grinds away," completely blind to any other factors: The teeter-totter trips even though the cat is already in the house, and the dishwasher washes even if someone has forgotten to put detergent into the cup.

In the future, however, we may expect to see some big changes from these old, inefficient ways. Future dishwashers will have a routine, but the routine will be altered from one load of dishes to the next by a microcomputer that "sizes up" the load. After the tub is loaded, the microcomputer will "decide" how much water is needed and at what temperature; it will add just the right amount of detergent, and then spray until the dishes are clean. Then it will rinse only until the detergent is washed away.

We might call a dishwasher of this kind a "smart" dishwasher. We will also see smart clothes washers and driers, smart ovens, refrigerators, food blenders, TV sets and radios, tape recorders, automobile brakes, lawn mowers, sprinkler systems, fire control systems, door locks, air conditioners, and so on, almost indefinitely. We can expect telephones that sort out and "remember" the dozen or so numbers that we call most frequently, and typewriters that know how to spell.

The key to all of these smart devices, of course, is the microcomputer. You might think that the microcomputer has pushed out the need for timing in machines, but this is not so. The need for timing has only moved from one spot to another; it has taken up lodging in the microcomputer. Now the computer steps to the rhythm of the computer clock, and the machine obediently follows the orders from the microcomputer. In the next chapter we'll see more clearly why clocks are essential to computers.

14

How Clocks Conduct the Computer Symphony

SOMEONE ONCE OBSERVED that if elephants ate and converted into energy as much food, per gram of body weight, as hummingbirds, all elephants would be roasted. The value of the observation is doubtful, but its reason is based in geometry. Both elephants and humming birds are roughly the shape of a sphere. The *volume* of a sphere depends upon its diameter multiplied by itself three times; its *area* depends upon its diameter multiplied by itself two times. This means that as the size of a sphere increases, its volume grows much faster than its area. The "area-to-volume" ratio of a humming bird is about 80 times that of an elephant.

The ability of an object—whether elephant or something else—to radiate energy, or heat, depends strongly on the exposed *area* of the subject. So if the elephant were converting food to energy at the same rate as the humming bird, the elephant would not have nearly enough area to radiate energy away from itself fast enough to keep its temperature below the roasting point.

Until the invention of the transistor about 30 years ago, computers were approaching the fate of the cooked elephant. The first computer contained some 18,000 vacuum tubes which, of course, had to be quite warm in order to operate.

An electronic circuit board with many "chips" replaces vacuum tube systems that used to occupy whole rooms.

Later, more sophisticated computers contained more and more tubes. And like the elephant, the computer's volume grew faster than its area. So as the numbers of tubes in computers increased, it became harder and harder to keep them cool. The time was approaching when the computer simply would be too hot to work, and this fact threatened the future of computers and the work they could do. Today, thanks to transistors and miniaturization, a small desk-top computer can now perform more functions in less time than an early vacuum-tube computer that filled three large rooms, with a fourth room for its cooling system.

But there's another important reason that computers should be as small as possible. The computer is like the inside of a beehive, with many bits of information moving from one place to another. If these locations are spread out over a large volume, the computer will be slow compared to a smaller computer doing the same work, even though the information travels at nearly the speed of light.

Today's fast computers perform tens of millions of operations in a single second. With so much going on, and the speed with which the information travels, it's obvious that the computer needs some kind of "conductor" to see that each task is performed at the right moment by the right part of the computer. The conductor also needs to provide the computer with a "beat" to synchronize the various tasks.

Let's look more closely at this "symphony of information," the computer. Just as every piece of music must have a "score," every job to be done by a computer must have a set of instructions called a *program*. And just as an orchestra can play anything from Bach to the Beatles, a computer's program can be readily changed, allowing it to perform a great variety of jobs.

The program is stored in a special part of the computer called the "memory." The memory consists of thousands and millions of tiny switches, which are either open or closed. If the switch is closed, it represents a "1," and if open, a "o." Or it could be the other way around, if we choose. On-off switches can be made from tiny transistors that can be packed by the thousands into an area less than 1/2 centimeter on a side.

In Chapter 11 we saw how to change numbers into strings of o's and 1's, or bits. In the same way, the *alphabet* can be divided in half, then in quarters, and so on, so that each letter of the alphabet has its own representation in bits. With this process, both numbers and words and phrases formed from letters can be stored in a computer memory, with the use of simple on-off switches. We see now why there is such a natural match between digital communication systems and computers; both are ideally suited to manipulate o's and 1's which can represent almost any kind of information.

"Data," or pieces of information, are also commonly stored in computers. An example of data might be the ages of 100 different individuals. Suppose we want the computer to find the average age of these 100 individuals. This function is carried out by a part of the computer called the "central processing unit," or CPU. To find the average age of 100 persons, the computer must first find the sum of their ages and then divide by 100. The CPU performs these tasks under the direction of the program instructions stored in its memory. After the answer has been found, the next program instruction may direct the CPU to send the answer to the memory for later use.

At this point the computer "knows" the answer, but *we* don't. We need some way to communicate with the computer and allow it to communicate with us. These tasks are carried out by the part of the computer called "input/output." An example of an input device is a "card reader" which "reads" numbers from punched cards into the computer memory via the CPU. Perhaps this is how the 100 persons' ages got into the computer in the first place. An output device could be an electronic typewriter which, under instruction from the program, types out the answer we've asked for.

We've skimmed over one important point: how does information get from one part of the computer to another? We might imagine that each part of the computer is connected by wire or cable to every other part. This would be similar to connecting every telephone in the country to every other phone by a direct wire. But this would be expensive and cumbersome, and it is not necessary, since we rarely wish to be connected to more than one phone at a time. The solution for the telephone company was to use a combination of subscriber loops, trunk lines, and switches. The advantage of this approach is that the same lines can connect many different pairs of telephones.

The same idea is used in computers. where the connecting lines are called "buses." As we might expect from

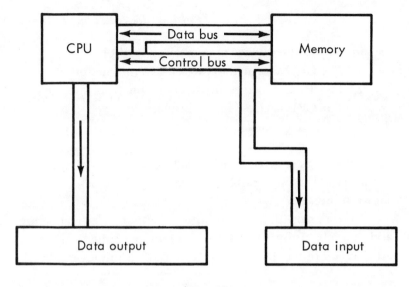

(Fig. 1)

the name, a bus is a line that carries many different kinds of information. Many computers have more than one bus. One bus may carry only data, and another may carry the signals controlling the sequence of operations in the computer, as shown in Fig. 1.

Now that we have explored the basic anatomy of a computer, we can appreciate the complexity and variety of tasks it can perform. But these tasks must be coordinated and synchronized with split-second accuracy. This is the job of the CPU, or conductor, whose "beat" is provided by a special computer clock. Unlike the orchestra, in which many musicians perform together, however, the computer carries out its tasks methodically, one step at a time. The pace for these steps is set by the computer clock, which generates a string of electrical pulses.

The clock pulse rate is determined largely by the speed with which the CPU can perform its duties and by the time it takes to store and retrieve information from memory. The fastest computers can perform 100 million operations per second. A typical sequence of operations might be:

1. The CPU asks memory to send an instruction.
2. The CPU receives this instruction from memory.
3. The CPU carries out the instruction.

These three steps make up what is called one complete "machine cycle" which is repeated over and over. Different parts of the cycle often require different lengths of time; so the parts of the cycle might be related to the clock pulses as shown in Fig. 2.

Considering the rate at which computers can generate and "recycle" information, who has enough time to listen to all that computers have to say? The human mind can absorb

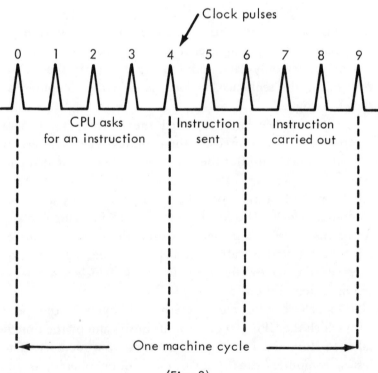

(Fig. 2)

information at the rate of only about 16 bits per second. So it appears that the only strong candidate is another computer.

Computers are often connected to each other by ordinary telephone lines, but when vast amounts of information must be sent almost error free, special lines, and even communication satellites, are used. As we learned in Chapter 12, digital Time Division Multiplexing is a powerful means for transferring information between locations at a high speed. And TDM is particularly well suited to computer "talk," since it is already in the form of o's and 1's. We learned also that TDM depends crucially on accurate time. So both the computer's ability to compute and its ability to communicate with other computers depend on very fine clocks.

15

Navigating Ships & Aircraft

AT NIGHT bats explore their surroundings by making short cries that are reflected back to their sensitive ears from surrounding objects. Nature has perfected this echolocation system to a remarkable degree. Some kinds of bats can distinguish between objects separated by only one or two centimeters, even while darting and circling about at full speed. Even more remarkable, the echoes may tell the bats something about the size, shape, direction, and motion of the object; so they can identify and catch particular kinds of flying prey on pitch dark nights. And contrary to popular superstition, they do not get caught in people's hair!

Only in modern times has humanity begun to use some of the sophisticated echolocation techniques that nature evolved for bats thousands of years ago. Most bat cries are of two kinds. The first is a simple tone, usually lasting 10 or more milliseconds, at a single frequency well above what human ears can hear. The second is a sliding tone, from high to low, much as a trombone changes its pitch when the slide is extended. This sliding tone normally lasts less than 5 milliseconds and helps the bat to determine the *distance* to objects; the single-frequency tone helps it to know the *speed* of its prey.

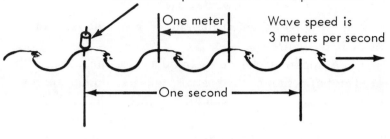

Cork bobs up and down 3 times per second

One meter

Wave speed is 3 meters per second

One second

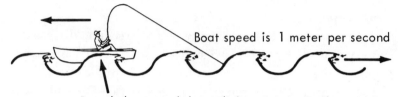

Boat speed is 1 meter per second

Boat bobs up and down 4 times per second

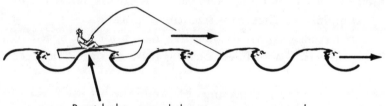

Boat bobs up and down once per second

How does the speed determining system work? Well, let's imagine a group of waves, all one meter long, moving along the surface of a lake at 3 meters per second. A stationary cork floating on the surface of the water bobs up and down 3 times a second, or at a frequency of 3 hertz. A small boat sailing directly into the waves at one meter per second bobs up and down 4 times a second—a frequency of 4 hertz; that is, people on the boat would see the wave crests moving by at a rate of 4 crests per second. And a boat sailing one meter per second *with* the waves bobs up and down at a frequency of 2 hertz; people on this boat would see the wave crests moving past them at the rate of only 2 per second, as shown in the figure.

This apparent change in frequency of the waves with speed and direction of the boat also applies to other kinds of waves, such as sound waves and radio waves. The pitch of the horn of an approaching car appears higher than it does for the same blast of the horn as the car moves away from the listener. The *amount* of the change in pitch is a measure of the car's *speed*.

This speed-frequency change effect was first discussed many years ago by an Austrian scientist, Johann Christian Doppler, after whom it was named "the Doppler effect." A Dutch scientist, Christopher Heinrich Dietrich Buys-Ballot, tested the constancy and reliability of the effect by persuading some trumpeters to blow their horns in an open railway car as it sped through the Dutch countryside. Since then our understanding of it has been greatly refined, and the Doppler effect has many scientific applications.

It has also shown us how bats determine the speed of flying objects. The echoes reaching their ears are "Doppler shifted" up or down, by an amount depending on the speed and direction of the reflecting object and the bat itself. Somehow the bat must measure the amount of Doppler shift to determine the object's speed; it is as though the bat can "remember" the frequency of its cry for comparison with the echo-signal frequency. Apparently the bat's "frequency standard" is imbedded in its echo processing system. But this is the subject of much research, and we are interested mainly in the fact that bats do carry some kind of frequency standard with them, and that we can use the same principles.

To determine the distance to objects, bats "measure" the time it takes the sliding-tone signal to travel to the object and back. The greater the distance to the object, the greater the round-trip delay. Thus nature has not only provided bats with frequency standards, but has also given them biological clocks to measure round-trip delay time.

Man's chief contribution to the system nature developed for bats was to build better clocks and couple them with stronger echo signals. Some of these are sound waves, or *sonar,* which means *sound navigation ranging* and describes a system used mainly to locate submarines. Others are radio waves, or *radar,* which is *radio detection and ranging,* used to probe the distant planets as well as to catch speeders on the highway. Whether sonar or radar, the principles are just the same as those used by bats.

Man has developed some variations on the bats' echolocation system, however, and most of these require some kind of time and frequency measurement. One variation requires the very best kinds of clocks man knows how to make, although the method of the system is very simple. Suppose you and a friend across town wish to find the distance between you. Each of you has a Super Space-Age watch synchronized with that of the other, and by agreement you send a pulse of radio signal at noon which arrives 9 microseconds later at your friend's location. Since your watches are synchronized—and since your friend knows you sent the signal at noon—his 9-microsecond measurement is precisely the travel time of the signal. Because radio waves travel about 1 kilometer in 3 microseconds, he knows you are 3 kilometers away.

This simple scheme is the basis for one kind of aircraft collision avoidance system that has been studied extensively. Of course, it requires that all participating aircraft carry the very best clocks, for even a 1-microsecond clock error translates into a 1/3-kilometer distance error.

Another kind of system does away with the need for carrying clocks precisely synchronized with the clocks at the broadcasting sites. But this advantage is gained at the expense of requiring the user to "listen" to signals from several locations. It works like this: suppose three transmitters are located at points of an equilateral triangle, as shown in Fig. 1. You are standing at the center of the triangle when all three

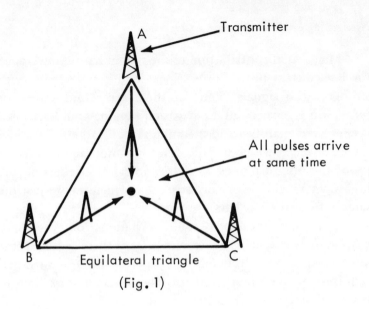

A — Transmitter

All pulses arrive
at same time

B   Equilateral triangle   C

(Fig. 1)

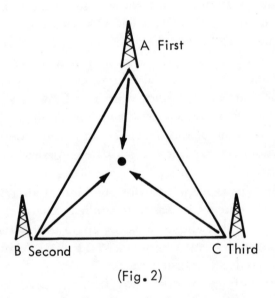

A First

B Second   C Third

(Fig. 2)

transmitters send a radio pulse at the same instant. Since you are at the center of the triangle, all three pulses arrive at the same instant at your location. The fact that they arrive at the same instant tells you that you *are* at the center of the triangle.

If you were standing closest to A, next closest to B, and farthest from C, then the pulses would arrive in the order A, B, and C, with the separations between pulses precisely related to your location. Using this method, you could always tell exactly where you were, anywhere within the boundaries of the triangle. To make the time-separation measurements, you must have a clock that measures time interval very accurately, but it need not be synchronized with the transmitter clocks. The transmitter clocks, of course, must be synchronized with each other.

Two of the most widely used navigation systems, Loran-C and Omega, operate in just this way. Loran-C—*Long Range Navigation*—stations transmit pulses at a 100-kilohertz frequency, and Omega stations transmit tones at several frequencies near 10 kilohertz. Loran-C stations provide position accuracy to better than one kilometer, but only within 1,600 kilometers or so of Loran-C stations. A few Omega stations provide worldwide signals, but the position accuracy is usually no better than a few kilometers. Navigators often need both Loran-C and Omega receivers to find their position accurately.

After the first satellites were launched, it occurred to scientists and engineers that the navigation transmitters did not have to be on the ground, but could be in satellites. This idea is the basis for a new satellite navigation system called the Global Positioning System (GPS), which is now being tested. Eventually 24 GPS satellites will circle the earth so that at any point on the globe, four satellites will be overhead. All 24 satellites will contain clocks that are precisely synchronized, to make certain that the signals are always sent at the correct moment and frequency.

As in the Loran-C and Omega systems, to find his position the navigator measures the difference in arrival time of signals from several different satellites, which continuously broadcast their location. One of the big advantages of GPS is that its signals are at a very high radio frequency, which is

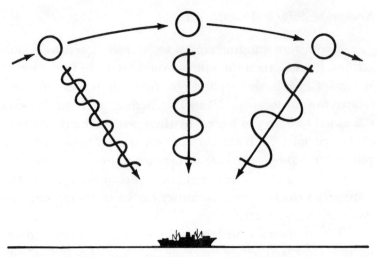

Frequency goes from high to low as satellite passes over

little affected by the earth's atmosphere. This is not the case for Omega or Loran-C. Their signals are stretched and distorted as they travel to the receiver.

Another satellite navigation system, called the Transit system, is used extensively today. In this system, the navigator needs to observe only the signals from *one* satellite. One single satellite, moving across the observer's sky, provides signals from not just one location, but from a continuous chain of locations. The navigator measures the Doppler shift—with which we are now quite familiar—of the signal as the satellite "rises" some distance from his location, passes overhead, and "sets," as shown in the diagram. As we know, this shifting frequency can be translated into changing distance to the satellite; and this information, along with the position of the satellite—which the satellite broadcasts continuously—tells the navigator what he needs to know to find his own position. With care, the navigator can average a number of Doppler-shift measurements to determine his position to better than 0.1 kilometer. Today many ships, submarines, and large commercial airplanes carry Transit receivers.

Scientists are working constantly to develop simpler, surer, and less expensive ways to keep track of ships and airplanes, and even trucks and buses on the highways. With more and more ships, aircraft, and vehicles moving about all the time, better systems for avoiding collisions are a growing necessity. All require better clocks, better use of what we know about time and frequency.

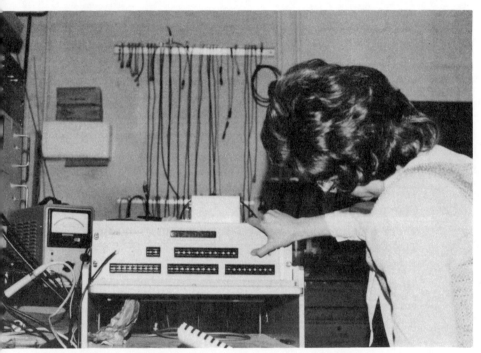

Tuning in a satellite receiver.

16

Tracking Nature

EARTHQUAKES, lightning, and violent storms such as tornadoes and hurricanes are among the most spectacular and sometimes destructive forces of nature. Each year storms and earthquakes kill thousands of people and destroy millions of dollars worth of property. These forces have always existed, but as populations increase and people continue to build more multimillion-dollar buildings and larger cities, with elaborate transportation and communication systems tying them together, the damages from "natural disasters" continue to grow.

Scientists are a long way from being able to *prevent* these disasters—although they keep studying possible ways to make them less devastating. Just being able to *predict* where and when a tornado or earthquake may strike could be a big help in saving thousands of lives and preventing some property damage. Yet such predictions are at present frustratingly unreliable; even simple day-to-day weather predictions are often far from accurate.

Satellites that can "see" the large picture of what's going on all around the world have helped. So have computers and electronic instruments that read and process weather information faster than human beings can do it. Meteorologists have a big responsibility in interpreting the information

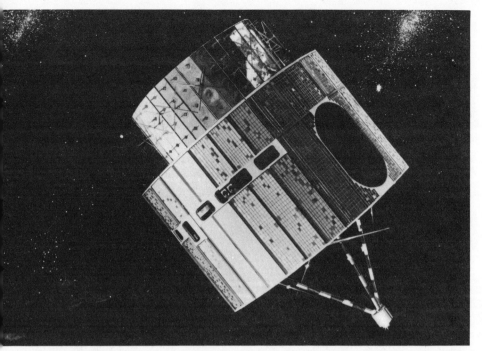

Geostationary Operational Environmental Satellites collect and transmit weather information. The satellite picture you see on your TV weather comes from a GOES.

and making forecasts. Ranchers move their livestock, stores change plans for sales, and school superintendents decide whether to close schools on the basis of the predictions. If a major storm is in the forecast, thousands of persons' plans and activities are affected. Armed with a better knowledge of what triggers the destructive forces of nature and how they build up, scientists can increase their ability to make more reliable forecasts.

Two important tools used in studying storms and earthquakes depend strongly on time and frequency techniques. One is radar, which we've just discussed as a tool of navigation. The other, oddly enough, depends on measurements involving certain kinds of stars called "radio stars," which emit radio signals.

We already know that radar signals are used to probe distance and speed of both nearby and far-away objects. Fortunately for weather studies, radar signals in certain frequency ranges easily penetrate thick fog and clouds, but are reflected by raindrops, hailstones, and snowflakes. Even more interesting, these reflections differ from each other; so observers can use radar very effectively both to study the motions of the particles in a storm and to determine what kind of particles are there. This kind of information is essential to developing any supportable theory about lightning and hail storms, and understanding the mechanisms of such storms.

Sometimes violent thunderstorms spawn tornadoes, which often develop so quickly that no one is aware of them until the dreaded dark, swirling funnel cloud is well underway. But there is growing evidence that before a tornado there's a kind of "pretornado" that can be spotted by special kinds of radar. At present little is understood about the pretornado or its possible place in tornado prediction, but it appears to be an important clue.

Radar is also used routinely to track a tornado once it was been spotted. With even a little information on the size, speed, and direction it is taking, weather observers can warn people in the path of the tornado to protect their property as best they can, get livestock to a safe place, and either take cover or flee to a safer area themselves.

Most violent storms are accompanied by dazzling displays of sharp lightning. Although scientists from at least as far back as Benjamin Franklin and his famous kite and key have been studying lightning, they still understand very little about it. Yet this understanding is more and more important. The National Oceanic and Atmospheric Administration says that lightning hits the earth somewhere 100 times every seecond, causing annual property damage of over $100 million a year and killing about 100 persons per year.

Since it's next to impossible to predict just when and where lightning *will* strike, the next best thing is knowing

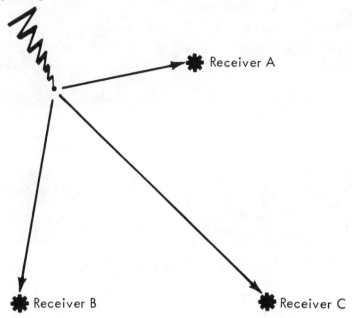

Lightning flash

Receiver A

Receiver B   Receiver C

Lightning flash is detected first at location A, then B, and finally C. Flashes at other locations will arrive at locations A, B, and C at other times.

when and where it *does* strike. The location of a flash of lightning can be found by using a scheme similar to the way the Loran-C navigation system works. We recall that a person can find his position by measuring the relative arrival times of radio signals from three different Loran-C stations.

To measure the location of a lightning flash, the situation is reversed. The three *transmitter* stations are replaced by three stations with radio *receivers,* which can detect the crash and crackle of radio noise that accompanies a lightning flash. The stations have synchronized clocks, so that the relative arrival time of the radio "crash" can be measured at the three stations. If the three receiving stations form an equilateral triangle, and if the lightning flash occurs at the center of this triangle, then the crash is heard at the same instant at all three stations. If the flash is at some other

A high-speed photograph of a lightning flash.

location, there will be different arrival times for the crash at each station, and this difference can be translated into the position of the lightning flash, as shown in the diagram.

High-speed photography, together with electronic timing techniques, is also important in the study of lightning. The eye is not quick enough to follow the details of a lightning flash; so what is really a rather complicated process appears to the eye to be a single flash that seems to change in brightness as it shifts position slightly. High-speed photography coupled with electronic timing of the radio crash reveals that lightning actually develops in several stages. First the lightning "searches out" a path for the flash to follow. This searching is done by what is called a "stepped leader," which makes its way from the cloud to the ground in steps. It moves about 50 meters in one microsecond, setting up a path for the current. It then pauses about 50 microseconds, and then moves on another 50 meters, finally reaching the ground in about 0.01 second.

After the conducting path has been established, the main lightning flash—the one we see—occurs; it lasts about 50 microseconds. Then after the flash there may be a number of other flashes at intervals of about 0.01 second. It's easy to see why the eye cannot follow this complex sequence of events, all of which take place in less than one second. Precisely timed instruments, however, can "see" what takes place, and so provide a way for scientists to study the mechanics of lightning. Although they may never be able to predict when or where lightning will strike, the more they understand about how it "works," the better prepared they are to cope with its effects.

If weather forecasting is risky, earthquake prediction is much riskier and can have much more far-reaching effects. A strong earthquake in a heavily populated area can be far more destructive to both life and property than even the worst tornado. If forecasters could predict with certainty when and where an earthquake would occur, thousands of lives could be saved by moving the people away from the area. Fire and flood damage could be largely prevented by shutting off gas, electric, and water lines. But the costs and efforts involved in these procedures would be tremendous, and if the earthquake didn't materialize, people would resent the disruption of their activities and would be reluctant to believe the next prediction—which might prove to be true. For these reasons, scientists and others working with earthquake studies are unlikely to make predictions until they are very sure that their predictions are reliable.

Today many of the countries of the world are spending considerable time and effort trying to understand the mechanisms of earthquakes. One of the most important activities is the collection of information about movements of the surface or "crust" of the earth. It may be that minor, almost undetectable movements of this crust give clues about the mechanisms of earthquakes, and possibly even signal the coming of earthquakes.

One of the most precise tools for measuring movements of the earth's crust involves the unlikely combination of clocks and radio stars. We can see stars at night because they give off light waves. But some stars also emit strong radio waves that can be received on special receivers. The "program" of most radio stars is very boring, and sounds like static on the radio. But this static provides a key ingredient in measuring the movement of the earth's crust.

To see how the system works, let's suppose that we have two antennas pointed at the same radio star, as shown in Fig. 1. One of these antennas is mounted on a truck, so that it can be driven to different locations. The other is mounted permanently, perhaps somewhere on the desert floor in an area being monitored and studied.

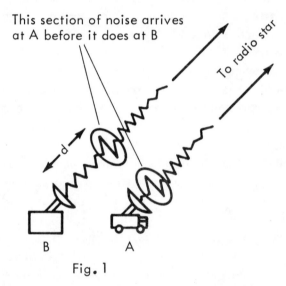

This section of noise arrives at A before it does at B

To radio star

Fig. 1

The section of radio noise has to travel an extra distance (d) before it reaches the antenna at B.

The radio star broadcasts a continuous stream of static or noise, which is also shown in the figure. The two noisy signals arriving at the two antennas are, of course, identical—in the same way that two persons hear the same program from a radio station although they are listening at different places. But we also see that because of the locations of the star and the antennas, a particular *section* of noise arrives at the

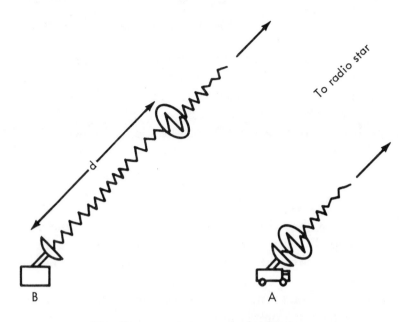

To radio star

B   A

The distance, d, is greater because
the truck has moved to a new location.

Fig. 2

Earthquake study equipment.

antenna on the truck at point A before it does at the desert floor antenna, at point B. If we have accurate clocks at the two antennas, we can measure precisely how much sooner the signal arrived at A than it did at B.

If we drive the truck to a new location, as shown in Fig. 2, the difference in arrival time of the two signals changes because of the new location of the truck with respect to the star and location B. We could also measure this new difference in arrival time very precisely.

Let's suppose now that the truck changes its position not because we drive it to a new spot, but because there is a slight shift in the earth the truck is resting on. This shift might happen because the truck and the desert antennas are on opposite sides of a "fault" line, or crack in the earth's surface along which the ground can slide in opposite directions. You may have seen pictures taken after an earthquake, where a house straddling the fault line has been split apart, leaving the two parts of the house separated.

Most of the radio star measurements of earth shifts are not nearly so dramatic as this. These visible shifts are easy to detect and measure. It is much harder to detect very slight movements, perhaps a shift of a few centimeters over a whole year, along a fault line out in the desert wasteland. Or the movement may be a slight uplifting of the desert floor. But these seemingly insignificant movements might be ominous signals that an earthquake is pending.

Before the development of accurate atomic clocks it was not possible to use radio stars to measure earth crust movement over large distances between antennas. But now such measurements are being made almost routinely. New approaches have also been developed using not radio stars, but satellites, and even a radio transmitter on the moon.

Nature often hides its secrets well, and some of its behavior seems impossible to predict. But time and clocks are helping man to understand some of the deep mysteries, and we shall discuss more of these in the last section of this book.

17

Sailing to the Planets

RIDING A BICYCLE seems as easy and natural to most cyclists as walking or breathing. But what if the bicycle had some sort of built-in delay, so that when the rider steered to the right, it might be several minutes before the front wheel changed direction? Or if the rider pedaled faster or put on the brakes, suppose there were several minutes delay before the bike responded.

This is a problem that faces the engineers and scientists who guide interplanetary space vehicles. The guiding signals travel as radio waves near the speed of light. But even at this high speed, the nearest planets are so far away that the signals take minutes to reach them. And the spacecraft itself travels so fast that its location has changed by many kilometers by the time the signal reaches it. Then it has changed again before the ground crew knows what has happened.

Let's suppose, for example, that a signal commands a space ship to change its speed slightly. After responding to the command, the vehicle returns a signal that tells of its success or failure. So it is twice the one-way delay time before the ground crew knows whether the command was carried out, and with what results.

The degree of correction needed, of course, depends on how much the vehicle strays from its intended path. But this can be determined only by knowing precisely where the vehicle is. Fifteen years ago it was not possible to tell, from the earth, the location of a distant spaceship, to better than a few thousand kilometers. Today locations are determined to a few tens of kilometers.

What made this 100-fold improvement possible? A number of things. One is that radar antennas are much larger now than they were 15 years ago. Another is that the sensitivity of radar radio receivers has increased greatly. But even with these improvements, the measurements could not be made without today's very stable clocks.

In Chapter 4 we learned how accurate clocks help sailors find their longitude at sea. But space ships are not on the surface of the earth; so longitude—at least as we normally think of it—is not very useful for space travel. What is important is knowing very accurately the distance, or *range,* of the space vehicle, and its speed. These measurements require measuring intervals of time within nanoseconds— one thousand-millionth of a second—over a period of several hours.

The basis for the *range* measurements is very simple. A radar transmitter sends out a sharp pulse of radio energy, which travels to the space vehicle; there it is transmitted back to earth. A clock at the radar transmitter tells when the signal is sent and when it returns. That is, the clock measures the round-trip time from the earth to the space ship and return. Since the signal travels near the speed of light, it is a simple matter to turn the time measurement into a distance measurement: Distance equals speed times travel time. Then to get the one-way distance, we divide the round-trip distance by two.

Because the range measurement is really a time-interval measurement, any error in the time measurement produces a corrsponding error in the range measurement. And the reason there is such a severe demand on the clock is that the speed of light is such a big number. Light travels about 30 centimeters in one nanosecond; so for every nanosecond of clock error during the measurement interval, there's a 30-centimeter error in the round-trip distance measurement.

We could determine the *speed* of the space ship by comparing several range measurements taken over a period of time. For example, suppose a particular range measurement shows a distance of 100,000 kilometers, and another measurement, one hour later, shows 101,000 kilometers. Assuming the vehicle is traveling directly away from the earth, we conclude that it is traveling 1,000 kilometers per hour.

But there is another way the speed can be measured, based on the Doppler effect, which we've met before. The Doppler effect has proved to be a very useful tool in many different problems of time and distance measurement. By now we're familiar enough with it to know that the amount of change in the pitch of an approaching or receding train whistle depends on how fast the train is moving. In the same way, a radio signal broadcast from a space vehicle moving rapidly away from the earth arrives on earth with a frequency lower than the broadcast frequency. The *amount* by which the frequency is lower is a measure of the "going away" *speed* of the vehicle.

One possible problem with this procedure is that the frequency broadcast by the vehicle may drift away from what we expected—perhaps to a lower frequency. What we interpret as an increase in speed of the vehicle is really a drift in the broadcast frequency. To overcome this possibility, a signal is sent at a known frequency from earth to the vehicle, where it is received and rebroadcast back to earth. Since the vehicle is moving away, the frequency it "sees" is lower than the one broadcast. So the vehicle rebroadcasts this lower frequency signal.

On the return trip, the signal is further Doppler-shifted down in frequency because of the movement of the vehicle. So the total frequency shift is two times what it would have

Two precisely synchronized antennas listen for signals from outer space.

been for a one-way trip. To get the correct speed, we must divide the Doppler frequency shift by two. The range measurements can be made with accuracies down to a few meters, and the Doppler shift measurements give the speed to about 4 ten-thousandths of a kilometer per hour.

Unfortunately, these high accuracies do not guarantee that we know the location and the speed of the vehicle to the same degree. You can understand this problem if you imagine that you are standing in the center of a circle one kilometer in diameter, and a friend is standing on the circumference, 1/2 kilometer away. A steel tape measure is stretched between you, and it reads 1/2 kilometer. If your friend moves directly away from you one meter, then the measure shows 1/2 kilometer plus one meter. But suppose that instead of moving *outward* away from you, he moves along the circumference of the circle one meter. The tape measure still reads 1/2 kilometer, even though your friend has changed his location.

Radar measurements are subject to this same kind of problem, since they measure distance, not direction. How can

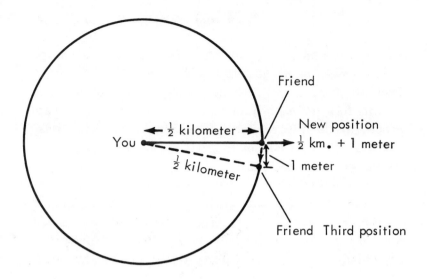

the problem be solved? Well, by measuring the distance to
the unknown location from two different places. Suppose *you*
move to a new spot—say 10 meters north of your old spot, as
shown in the figure. Then you measure the distance to your
friend from your new location. Now you know the old
distance—½ kilometer, the new distance, and the distance and
direction in which you walked. These three distances make
up the sides of a triangle, and there is only one triangle that
has sides with precisely these three lengths. So knowing the
dimensions of this triangle, you can find the direction to your
friend.

Obviously, it is not practical to move huge radar
antennas from one location to another. But if there are two
or more radar antennas at different locations, with perfectly
synchronized clocks, these can serve the same purpose as
moving just one antenna to a different location. However,
when there is only one antenna, it is still possible to make the
measurements, for the movement of the antenna is performed by the earth itself; as it rotates about its axis, it moves
the antenna from one point in space to another.

In spite of the great accuracy of radar measurements and
man's knowledge of the paths followed by the planets, how-

ever, the techniques we have described cannot provide the needed precision for journeys to the most distant planets. Ironically, for these trips man must return to the most ancient form of navigation—navigation by the stars. Space probes to the planets carry television systems that relay pictures of the surface of the planet back to earth. These systems also send back images of the background stars. Since the positions of the stars are well known, it is possible, using a series of such images made as the space ship moves along, to determine its location very precisely. Thus man's earliest form of navigation coupled with one of his latest inventions, television, allows him to journey to the planets—indirectly now, with space probes, but in the future, perhaps, in person.

III

Time
&
Science

18

When Did It All Start?

To ASTRONOMERS, the planets in our solar system are practically in our own back yard. But we know that our solar system is only a tiny part of one galaxy, the Milky Way; and our galaxy is only a small part of the universe. Time and time measurement play a very important part in scientists' efforts to understand the universe. How did it begin, and how will it end? What was before the beginning, and what will follow after the end?

Questions like these go right to the heart of the meaning of time. Is time an invention of man, or does it exist by itself, as Sir Isaac Newton believed? If time is an invention of man, then perhaps questions about beginnings and endings are only "apparent" questions rising out of the way people use words. We cannot answer these questions finally, but we can sketch man's present concept of the birth and evolution of the material universe.

The universe probably burst into being some 10 to 20 billion years ago—and its future course was set in less time than it takes to read this chapter. Before 1965 few scientists would have made this bold statement. Although there had been theories about how the universe began, there were not enough observations or "hard evidence" to support the theories or make them worth serious consideration.

Dr. Arno A. Penzias and Dr. Robert W. Wilson discoverers of cosmic background radiation.

Then in 1964 two radio astronomers in New Jersey, Arno A. Penzias and Robert W. Wilson, observed a puzzling, faint radio noise. Their studies of this noise later proved to be so important that they shared the 1978 Nobel prize in physics for their work. Their studies soon changed speculation about the history of the universe from fanciful theories into serious investigation. We shall discuss their observations in more detail, but first we must look at what little was known at the time they began their studies.

One earlier and very important observation had been made in 1929 by a lawyer-turned-astronomer, Edwin Hubble. Hubble discovered that wherever he pointed his telescope, the stars seemed to be rushing away. It was as though there had been some gigantic explosion in the distant past which had flung apart the universe. And it was still expanding, with its parts all moving away from each other, in a way similar to dots painted on the surface of an expanding balloon.

If we measured the directions and speed of these spreading dots, then we could estimate the *time* when the balloon started to expand. So it is with the universe. As we gather information about the distances and speed of stars scattered throughout the universe, we can reach backward in time, with science and mathematics, and create a "movie" of its evolution.

Of course we know that when we look at distant stars today, what we are seeing is what happened thousands of

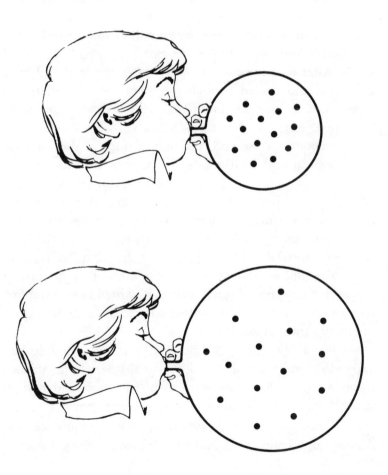

years ago, because it takes light thousands of years to travel over vast distances that separate us from the remote stars. If we run our movie backward, we find that the "show" started 10 to 20 billion years ago. As the picture on the screen retreats in time, we see stars and galaxies approaching each other, crowding ever closer together, until all the mass of the universe is concentrated in one single point at the time of the explosion, or the "Big Bang," as physicist George Gamow named it.

What can we say about the energy and matter of the universe at such very high densities? Not much before the first 0.01 second or so after the explosion. During that brief period of time the pressures and temperature would have been so great that only the forces holding atoms together are important, and these forces are not very well understood today. After 0.01 second, however, the universe would have expanded and cooled enough for other, better understood forces and actions to dominate. Let's see what probably happened.

At about 0.01 second, the temperature of the universe was near a hundred billion degrees Celsius, and it consisted mostly of photons of light, electrons, positrons—which are like electrons but with a positive charge—neutrinos and antineutrinos—which are small, massless particles—and lesser amounts of protons and neutrons. All of these particles interacted vigorously with each other; so the "soup" of the universe was thoroughly mixed. This is important because it means that any special conditions that existed before the first 0.01 second had been "smoothed out" and would not affect the future train of events.

After a few hundred thousand years, the universe had expanded and cooled to the point where the various components could combine to form hydrogen and helium gases. Eventually, because of the gravitational forces that everything exerted on everything else, clumps of gas condensed, forming stars and islands of stars, the galaxies.

In the mixed, "infant" universe, the densities were so great that photons of light continually collided with each other and with other matter, so that they could not escape. As the universe cooled to about 3,000 degrees Celsius, the photons of light reacted very little with each other and with other matter. They simply escaped into the universe as it expanded. So if the Big Bang theory is correct, this "antique" radiation that existed at the very beginning should still exist today—but in a very dilute form. The theory predicted that the most intense radiation would be concentrated in the microwave region of the radio spectrum, and would have nearly equal intensity in all directions.

In 1964, Penzias and Wilson were testing a new antenna at the Bell Telephone Laboratories in Holmdel, New Jersey. They had taken every precaution to make the antenna as sensitive as possible, and they were concerned when they discovered a faint radiation at 4,080 megahertz, in the microwave region, which they could not get rid of, no matter how hard they tried. Furthermore, this signal seemed to be independent of the time of day or season of the year.

Word of their problem reached several scientists who knew about the Big Bang theory. They immediately suggested that the "delinquent" radio signal was the "fossil" radiation left over from the Big Bang, and that further observations should be made. Since then, many observations at other frequencies have shown a variation in intensity of the signal with frequency, which had been predicted by the Big Bang theory. And so the theory has become generally accepted today.

Will the universe continue to expand forever? Today no one is sure. The answer depends on the density of matter in the universe. If the density is above a certain critical amount, the universe will eventually stop expanding and fall back upon itself. If the density is below this critical amount, it will expand forever.

The reason for these two possibilities results from the

struggle between the first explosive outward-slinging of the universe at the time of the Big Bang, and the opposing gravitational attraction between the pieces of matter in the universe. The situation is a little like rolling a ball up a hill. If one throws the ball hard enough, it will roll to the top of the hill and over the other side. But with a weaker throw, it will roll part way up, stop, and roll back again. The Big Bang theory predicts the critical density, but not enough observations have been made to tell whether the density of the universe is above or below this critical level.

So *does* the universe have a beginning and an end? If the matter in the universe is below the critical density, then perhaps the universe continually contracts and expands from "big bang" to "big bang." But the question may also be meaningless. Perhaps we are looking at things in the wrong way. In earlier times, people believed that the earth was flat, and that if anyone walked or sailed far enough, he would fall over the edge. Or perhaps the earth was flat and extended infinitely far out in all directions, and one could walk or sail outward forever.

This concept of a flat earth produced many troublesome questions. If the earth was flat and finite, what was beyond the edge? If it was flat and infinite, how could one tell where the finite ended and the infinite began? Of course these bothersome problems arose out of a faulty model of the earth that people generally accepted as correct. The problems vanish when we accept a spherical earth.

Perhaps as we gain a better understanding of the universe, the paradoxical nature of time will vanish also. Perhaps time does not extend from the infinite past into the infinite future like a line with no ends. Maybe it is more like a line closed upon itself in the shape of a circle. It may just *appear* to be endless, in the same way that a very short length of the arc of a very large circle appears to be straight. We label events on this short length of arc as "before," "now," and "after." But if these events lie on a circle, the situation is

somewhat like that of three runners looping many times around a track in a long-distance race. The runner who appears to be in front may in fact be many laps behind. The apparent endlessness of time is only one of its many puzzles.

19

The Arrow of Time

WE'VE NOTED that one of the most obvious, commonsense characteristics of time is that, whether it moves in a "straight line" or a "circle," it seems to flow in just one direction. We can explain this idea more fully by imagining that we are watching a simple movie. The scene is a school science room, and the opening shot shows the teacher walking across the front of the room to a table with a sink. On the wall behind the table is an old-fashioned clock with a swinging pendulum and a second hand that jerks forward with each one-second swing. The teacher picks up a glass jar from the table, fills it half full of water from the faucet at the sink, and sets it down again for a few moments while he speaks to the class. He then picks up an eye-dropper filled with blue ink.

At this point the camera zooms in for a close view of just the teacher's hand and the glass jar as he carefully drops one drop of ink on the surface of the water, then withdraws his hand from the camera view. Slowly the ink spreads through the water, until all of the liquid is a uniform pale blue. As the camera runs on for another 20 seconds, nothing changes. The only movement is the swinging pendulum of the clock; because the camera is in so close, the face of the clock is cut out of the picture, and we see just the pendulum.

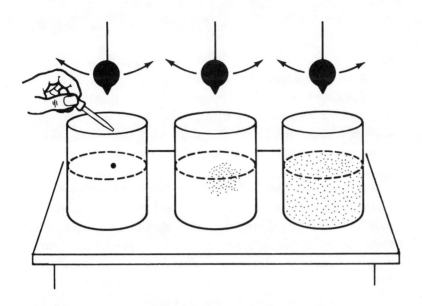

Now let's stop the projector and cut out the piece of film we have just seen. With scissors we cut it into hundreds of pieces, each piece being just one frame of the film. Next we dump the frames into a bag and shake them up. Then we ask: Can we remove the pieces from the bag and glue them back into their original order?

At first we're probably overwhelmed by the size of the task; but if we think about it for a moment we realize that, although it would be tedious, we can probably put at least some of the frames back into their correct order. We know, for instance, that we must arrange those from the first part of the film so that they show the teacher crossing the room, drawing the water from the tap, and dropping the ink into it. Another order would not produce a sensible scene.

But can we always cut a film into many frames and then reassemble them? Let's think about the last 20 seconds of the film, where the ink has dispersed uniformly into the water and the only movement is the swinging pendulum. Can we take this piece of film, cut it into frames and mix them up, and then arrange them in a sensible order? Probably, because each successive frame must show the pendulum slightly displaced from its earlier position.

Let's suppose we have glued the frames together so that we have a sequence showing the pendulum swinging smoothly back and forth. But then we find there's another problem. Although we're sure that we can assemble the frames into an order that makes sense, we cannot be certain that it is the *original* order. This is because if we assembled them in *reverse* order, we would still have a strip of film showing the swinging pendulum in an otherwise motionless scene. In other words, running this piece of film forward or backward makes no difference. Both show the uniformly blue water in a jar, with a swinging pendulum behind it.

We might say that during this part of the movie, *time* can run either forward or backward. This is not the case for the parts of the film that show the teacher filling the jar and dropping ink into it, and the ink spreading into the water. Water does not leap from a jar back into a faucet, and ink does not separate itself from water and return to a dropper, leaving the water clear. The second hand on a clock does not move backward.

It appears that our feeling about the direction of time is closely related to whether a series of events seems likely or not. Perhaps time does not have a direction in a universe that consists of nothing but swinging pendulums—or in one where there is no motion at all. But the "real" universe is teeming with activity and is much more complex than a group of swinging pendulums. It consists of a variety of objects interacting with other objects, producing new circumstances that probably have not occurred before and are not likely to be repeated in the future.

Let's return to our movie again and look more closely at the very end, the part that shows only the jar of blue liquid on the table. The camera moves in very close now, so that just the jar fills the whole frame and we can no longer see even the swinging pendulum. All we see is 15 seconds of a motionless, unchanging image. The ink has dispersed evenly in the water, and we see no movement at all.

Suppose we take this segment of film, cut it up, mix the frames, and try to put them back together. This seems an easy task. Each frame looks like any other—just a uniform blue background. It doesn't make any difference in what order we line up the pieces; we get the same image on the screen. For this part of the movie, time can run forward, backward, or even in random order. We might say that time doesn't even exist.

But suppose we use a powerful magnifying glass, so that we can follow the path of an individual particle of blue pigment in the "sea of blue." We see the particle moving

through a thicket of other particles, occasionally colliding with another one. At the instant of the collision, the particles bump and move apart in new directions, like two billiard balls colliding on a billiard table.

If we run this bit of film backward, we see a similar scene—particles moving past each other, particles occasionally colliding and moving apart in new directions. Whether the film runs forward or backward, we see individual particles move as we would expect them to.

Apparently the laws of nature that govern the motion of particles are not affected by the direction of time. We can put "forward" time or "reverse" time into these laws and we always see something that is sensible. If we reverse the time in the law governing the motion of a satellite around the earth, all that happens is that the satellite revolves around the earth in the opposite direction. This is certainly possible, since we can launch a satellite in any direction we wish.

It seems to be *circumstances,* then, and not the basic laws of nature, that dictate the directional flow of time. It is the circumstance that the ink was first "bottled up" in the dropper, and then released into the water, that gives time a direction. After the ink has completely dispersed throughout the water, there is no longer any sense of time, although the ink particles continue to move according to nature's laws of motion.

Just what is it, then, that is distinctive about the circumstances that lead to a direction for the "arrow" of time? The important point seems to be that any thing or event that has "time directedness" must at some point be *organized.* We might say that the ink particles, before they were dropped into the water, were organized into a small volume of space inside the drop of ink; and after the ink was dropped into the water they became disorganized by flowing randomly among the molecules of water.

This flow from organization to disorganization, which scientists call *entropy,* seems to be a natural feature of life and of the universe. People and animals grow old, automobiles rust, rooms go from orderly to messy, stars grow cold. Someday the universe may contain nothing but burned out stars. Then it will be like our jar of pale blue liquid, where the direction of time has no meaning.

Is there anything that works *against* the natural tendency of the universe to go from order to disorder—anything that can *reverse* the direction of time, so to speak? Certainly the inventive mind and hands of man seem to slow—and even at times to reverse—the march from order to chaos. But even in this there is a catch. Let's look at just one complex and "organized" creation of man—the automobile.

An automobile consists of many parts, made from a wide variety of materials. The engine is metal—iron or steel or aluminum; the tires are rubber and various fibers; the windows are glass, and the seats are probably covered with plastic. All of these materials were, at some earlier stage, in a much more primitive or disorganized form. The iron was imbedded in iron ore. The plastic was probably made from oil, and the glass from sand. So it would seem that when men make automobiles they make something that is organized out of what was formerly disorganized.

But it takes a lot of *energy* to refine the raw materials needed to make an automobile, and it also takes energy to assemble the parts. This energy may have come from burning coal and natural gas. Much of this energy goes into making the car, but some of it also escapes unused, into the atmosphere; some escapes as heat up steel mill chimneys, for instance, and some as heat radiating from cooling engine blocks that have just come from their molds. This energy is forever lost to us, since there is no reasonable way we can harness it, once it has escaped into the surrounding environment.

So the price for creating order out of chaos is a large expenditure of energy, only part of which is used to create the new order. Put another way, the organized *energy* in the form of coal and gas has been exchanged for organized *matter,* but the amount of organization created does not equal the amount spent. It's as though we're trying to run up a down escalator.

In the end, the ever downward flow of the "escalator of the universe" is probably traceable to the Big Bang that started it all. At that instant the universe was tightly organized into a point in space; and ever since then, the matter and energy of the universe have been expanding. Primitive peoples—and civilized men throughout the ages—have thought of the universe as the model of order. The stars and the planets, the sun and the moon could be depended upon, century after century, to hold their positions and courses in perfect order. And from the viewpoint of humankind, this sense of order is valid.

But humankind has not been here very long. If we think of the whole history of the universe as lasting one day, men and women have been on earth for only the last few seconds of that day. The flow from order to disorder in the universe is so very slow, spreading over many billions of years, that man can barely perceive it. Yet it is this slow change from the initial order just before the Big Bang to the "orderly disorder" that has followed it that seems to provide the ultimate direction for the "arrow" of time—and the generally accepted belief that time flows in only one direction.

20

Is a Minute Really a Minute?

THE WAY WE LOOK at things has a great deal to do with what we see. And our ability to "look" with our thinking at things we cannot see with our eyes is one of humanity's most amazing capabilities. This kind of abstract thinking is basic to scientific discovery and to recent developments in physics that involve time.

To help us understand this abstract world of modern physics, let's do a kind of warmup exercise. Suppose that quite unexpectedly, and for the first time in your life, you saw a bat. And suppose further that you had never read or heard that such a creature even existed. What would you think it was?

At first you'd probably believe you had seen a mouse with wings, or perhaps a bird with a mouse's body. It seemed to be both mouse and bird at the same time, but more like one than the other, depending on whether it was flying or hanging from a tree limb. Later, of course, you would find that it was neither bird nor mouse, but a bat—something you had never seen before.

Ideas like the direction of time, concepts of space and motion, seem very familiar to us, since we have clocks and watches and can move about in space. It's very tempting to project these ideas into areas we *cannot* experience directly. We cannot see atoms, for instance, and we cannot travel to even the nearest star beyond our sun. We see atoms only indirectly, as they are revealed by atom-smashing machines. And we examine the stars only through telescopes that magnify their images.

Perhaps we should not be surprised if our notions of time, space, and motion do not quite fit the very small and the very large. When we try to understand time and space as these ideas apply to the small and the large, we *begin* with our old ideas, just as you might try to describe the bat in terms of birds and mice. But if we are to understand the real nature of time, we must be ready to give up our old concepts and accept time for what it is, rather than for what it appears to be. If we don't cling to the old explanations, then certain mysteries will disappear—just as the mysteries of a winged mouse disappear as soon as we realize that the newly discovered creature is not a mouse at all, but a bat.

With that background, let's think about a problem that challenged scientists like Sir Isaac Newton and Dr. Albert Einstein: Is time the same for everyone everywhere? Would two identical clocks always keep the same time, no matter where they were? Would it make a difference if one was racing through space aboard a space ship? Common sense tells us that a second or a minute or an hour is the same for everybody, everywhere, just as a centimeter or a degree of temperature seems to be the same wherever we go. This feeling is usually expressed by saying that time is "absolute."

We can illustrate what the term *absolute* means by thinking about a tall tree. If we stand near the base of the tree looking up into its branches, it appears to be very tall. From a distant hill, the same tree appears small. Yet we don't believe that the height of the tree depends on our distance from it. We know that it has a true, or *absolute,* height that has nothing to do with whether we view it from nearby, or from a moving car, or from a distant hill.

In the same way, we believe that a length of time—one minute, let's say—does not depend upon our motion or location in space. A minute is a minute. But is it? Surely common sense says that it is, that it must be. But by now we may be beginning to suspect common sense just a little. And as we shall see, our everyday ideas about absolute time and distance are only illusions that arise because we're used to living in a "snail's pace" world. If we were used to moving at speeds near the speed of light, then our ideas about time would be different.

In 1905, Albert Einstein startled the scientific world with his famous "Special Theory of Relativity." According to Einstein, our commonsense ideas about space and time are not correct, although the errors cannot be noticed as long as we are moving at slow speeds or observing objects that are moving at slow speeds. Light, however, moves very fast. In fact, according to Einstein, nothing can go faster than light.

Let's explore this idea a little more. We'll imagine that we are standing at night on a roadside looking toward the headlights of an oncoming car. If the car is moving toward us at 100 kilometers per hour, at what speed does the light from the headlights approach us?

We might reason: Well, the light leaves the headlights at the speed of light, and the headlights are traveling with the car at 100 kilometers per hour. So the light approaches us at the speed of 100 kilometers per hour plus the speed of light. But according to Einstein this conclusion is impossible. No measurement can show a speed greater than the speed of light. And the fact is that if we measure the speed of the light from the headlights of the oncoming car, we would find it traveling at just the speed of light, not the speed of light plus the speed of the car. This result leads to some amazing conclusions about the nature of time.

To explore these conclusions a bit, we'll conduct what Einstein called a "thought experiment." Perhaps this will help us understand why time is not absolute, as common sense tells us, but is *relative*.

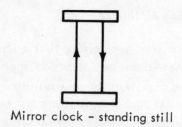

Mirror clock – standing still

Fig. 1

Car moving in
this direction

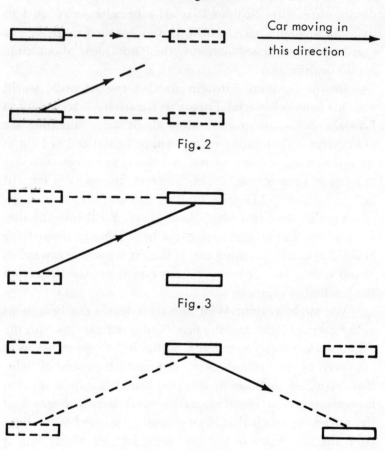

Fig. 2

Fig. 3

Fig. 4

We'll imagine that we and the man in the car have identical clocks of a very simple but special and unusual design. Each clock consists of two mirrors, facing each other but separated by a small space, as shown in Fig. 1. One "tick" of the clock is the time it takes one pulse of light to reflect back and forth once. Since the two clocks are exactly alike, they tick at exactly the same rate. Or do they?

Let's suppose that we can see the clock in the man's car, and can see the light pulse bouncing back and forth—which of course we cannot, for all of this happens much too fast. We'll imagine it in very slow motion. At some instant of time we see a pulse leave the bottom mirror of his clock and start toward the upper mirror, as shown in Fig. 2. As the pulse moves toward the upper mirror, the whole clock, of course, is moving with the car; so by the time the pulse reaches the upper mirror, it is in a new position, as shown in Fig. 3. Similarly, on the pulse's return trip the whole clock is in still another location, as shown in Fig. 4.

We can see from this series of figures that the light pulse in the car takes a zigzag path, instead of the straight, up-and-down path shown by *our* clock, which is standing still. So this zigzag path for one tick of the clock in the car is longer than the path taken by the pulse of light in our clock for one tick.

Furthermore, even though the clock is moving with the car, its pulses still travel with just the speed of light, as stated by Einstein, and not faster than the speed of light. This constancy of the speed of light combined with the zigzag path means that the clock in the moving car appears to us to be ticking slower because its pulses must cover a greater distance for one tick. So we conclude that the clock in the car runs slower than our clock.

But what does the man in the car see? If he looks at his own clock, he sees the pulse of light bouncing back and forth in just the same way that we see our own clock—because his clock is not moving with respect to *him;* he is moving right along with it, at the same rate. But when he looks at *our* clock, he sees *it* moving by at 100 kilometers per hour, relative to himself, with the light pulse taking a zigzag path like the one *we* see when we look at *his* clock. So he naturally concludes that it is *our* clock that is running slow.

Apparently time kept by identical clocks is not the same for observers who are moving with respect to each other. Or we might say that time is not absolute, as common sense tells us it is; it is *relative*.

As we have said, life on earth moves at a very slow pace compared to the speed of light; so we are not aware of the relative nature of time. But intelligent creatures on some other planet, where life moves very fast, would have everyday experiences that showed them that time is not absolute. They would probably be surprised to learn that we ever believed that it is or could be absolute.

In the world of relativity, the same event appears to occur at different times for different observers who are moving relative to each other. There is really no such thing as an event happening at the same instant—or with absolute simultaneity—for all observers.

21

Time Comes to a Stop

THE SPECIAL THEORY of Relativity was one of the most daring and revolutionary theories of all time, but Einstein thought there was more to explore. Special Relativity said that time and motion are related, and Einstein wondered whether there were other relationships, such as between gravity and changing or accelerated motion, and what impact these relationships might have on time.

Let's think first about the question of gravity and accelerated motion. Suppose that an astronaut is in a spaceship that is falling freely toward the earth, gaining speed all the time. The astronaut, although surrounded by the spaceship, is also falling freely—just as he would fall if he were outside of it. So from his point of view, he is just floating inside the spaceship's cabin, in the same way that he would if he were in outer space, far from the gravitational pull of any of the planets or stars. Because he is moving at the same speed as the spaceship, he always keeps his same position inside of it. If the spaceship has no windows and all he can see is the cabin around him, he might suppose that he and the spaceship are, in fact, simply drifting in outer space, free from the effects of gravity.

At this point in such a supposition, Einstein made

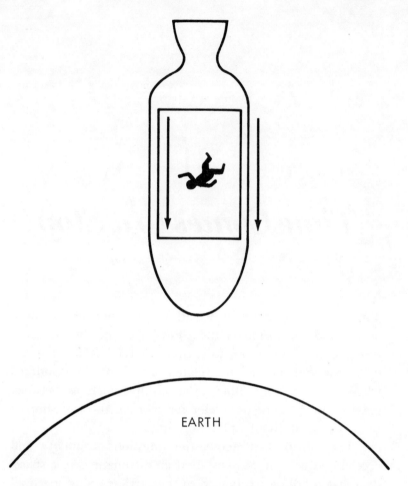

EARTH

another daring suggestion. He said that there is no possible way for the astronaut to tell which is true! Unless he can see something outside his spaceship to *relate* to, he cannot tell how fast he is moving, or even whether he is moving at all. And even then, he can tell only what his movement is *relative* to the object he is observing.

The big difference between Einstein's earlier suggestion and this new one is that in the first one, the observers were moving with *constant* speed with respect to each other, but his new suggestion is concerned with observers who are *changing* speed, or *accelerating*. This new suggestion was part of what is called Einstein's "General Theory of Relativity." It is more general than "Special Relativity" because it includes both uniformly moving observers and accelerated observers.

Let's do another thought experiment to help us understand what Einstein meant, and what effects his theory has on measuring time. We'll imagine that you are riding on the outside edge of a very large and fast-moving merry-go-round. The operator sits at the center of the merry-go-round, and the woman who takes tickets stands on the ground near the edge, as shown in the diagram. Each of you has a Super-Special Space-Age watch, and all three watches run at the same rate when lying next to each other on the ground.

As the merry-go-round spins, you feel a force, the *centrifugal* force, which seems to hurl you away from the merry-go-round. The operator does not feel this force because he sits at the center. This is the same principle that makes the

ice skater at the end of a "crack the whip" chain feel a much stronger force than the skater feels at the head of the chain.

What does Special Relativity tell us about the relative rates of your watch, the ticket collector's watch, and the operator's watch? First we notice that the operator and the ticket collector do not move toward or away from each other; they keep their same positions with respect to the ground and to each other. So according to Special Relativity, *their* two watches should tick at the same rate, as they see the situation.

But what about your watch and the *ticket collector's* watch? As we learned from the previous chapter, the ticket collector would say that your watch was running at a different rate from hers because your watch is moving with the merry-go-round.

Your watch and the *operator's* watch present a new situation. They are both on the spinning platform, but the distance between them does not change. So you would say that according to Special Relativity your watch and the operator's watch should tick at the same rate. But the operator notices that this is not true. Your watch seems to be ticking more slowly than his. The only apparent difference is that there is a centrifugal force at your watch, which is not present at the center of the merry-go-round. So the operator concludes that clocks tick more slowly where the force is stronger. Some experiments would show him also that the faster the merry-go-round spins and the greater the centrifugal force as you move around the curve, the slower your watch ticks relative to his.

What has this to do with clocks and the pull of gravity? Well, let's suppose that as you ride the merry-go-round you are blindfolded. You still feel the force that seems to fling you outward away from the merry-go-round but there's another possible explanation for it. Suppose you discover that the merry-go-round has stopped but you still feel the same force. You look around and find that directly behind you, resting

on the ground, is a massive chunk of extremely dense material—the kind of matter we might find in a burned-out star. What you feel is the *gravitational* pull of this very dense material.

According to Einstein, as we saw in the spaceship example, the gravitational explanation is just as good as the spinning-merry-go-round explanation. Both produce a force; and if one is blindfolded or in a spaceship with no windows, there is no way to tell which is responsible for the pull. Whether an object is flung away by one spinning body or pulled toward another body, the effect is just the same. So we conclude that clocks run more slowly where the gravitational attraction is stronger.

This raises an interesting question: Is there any object that could produce a gravitational pull so strong that a clock in its neighborhood would stop ticking completely—and time would come to a stop? There is growing evidence that such objects do exist. They are called "black holes" and they were predicted by the General Theory of Relativity long before anyone thought they might actually occur in nature.

What is a black hole? It is the remains of a very massive burned-out star. The energy of stars is produced by a continuous, nuclear bomb-like explosion at the star's center. Although it takes millions and millions of years, the fuel feeding the nuclear fire eventually burns up, and the star begins to cool and to collapse under its own weight. If the star is massive enough, the collapsing continues until all of the mass of the star shrinks to just a point in space.

The gravitational pull near such an object is so great that a nearby clock would stop ticking, and even light could not escape. This is why these objects are called black holes. A black hole cannot emit light, and any object that is nearby is pulled into it like water disappearing down a drain.

If it does not emit or reflect light, how can it be seen? Only indirectly, by the effect it produces on nearby objects, such as stars. Certain peculiarities in the radiation from space in the vicinity of some stars suggest that black holes really do exist.

If these observations are confirmed, what do they mean? Will all matter eventually be swept into one gigantic black hole? No one knows. It seems that each step we take toward new understanding of the universe creates more problems than it solves, as we have observed before.

We have come a long way from our commonsense ideas about absolute time. We have found that time is relative and depends upon gravity and motion, both uniform and accelerated. One may wonder whether Einstein's theories are the last word, or whether some new theory may come along to replace them. It may very well be that some new observation will be made that does not agree with the Relativity Theory. Then the Relativity Theory will have to be modified and perhaps even replaced. But if this happens, the new theory will have to agree with Relativity in the areas where both apply, just as Relativity Theory agrees with our normal, commonsense ideas of space, time, and gravity where speeds and forces are low.

Each new theory about nature stands on the shoulders of the theory it replaces. But the new theories seem to be more abstract and more distant from our commonsense ideas about nature and the universe and what our physical senses tell us. This change from a commonsense understanding of nature to a more abstract concept was greatly speeded up early in this century with the introduction of Relativity, and with another new idea, the Quantum Theory of matter, which we shall look at next.

22

Blurred Time

WE'VE FOUND that gravity and speeds near the speed of light do some strange things to time and clocks. If we could measure time more precisely, could we perhaps understand these things better? How precisely *can* we measure time? Is there anything in nature that keeps us from measuring time perfectly?

As a matter of fact, there is. This surprising answer evolved out of a new theory about matter called the *Quantum Theory,* which deals with the microworld of atoms and electrons. Just as we found that we could not apply our everyday, commonsense notions of time when speeds are high and gravitational forces are strong, we shall see that our everyday concepts about time and matter do not apply when we study the very small—atoms, electrons, and other particles that are the building blocks of the material universe.

One of the important ideas of Quantum Theory is very old: Whenever you make a measurement, the measurement process itself disturbs the thing you are trying to measure. Suppose you wanted to measure the temperature of water in a glass you have just filled from the kitchen faucet, for example. This seems easy enough. You just put a thermometer into the water and wait a few minutes for the thermometer temperature to settle at some value.

But suppose you want to measure the water temperature with extreme accuracy. Unless you are very lucky, the water will be at one temperature and the thermometer at another—let's say higher—temperature. When you put the thermometer into the water, it raises the water temperature ever so slightly, so that the final water temperature is a tiny bit warmer than it was before the thermometer was inserted. In other words, the measurement itself has changed the temperature of the water.

Of course if you wanted to measure the temperature of the water in a swimming pool, inserting the thermometer would have practically no effect on the temperature. But if you wanted to measure the temperature of water in a very tiny glass—say about the size of a thimble—then inserting the thermometer might produce a change that you could not ignore. We begin to see now why the measurement process itself can interfere with the measurement—particularly if we are dealing with very small objects.

We have the same kind of problem when we measure very small bits of time. Suppose we have a super-clock—even more accurate than an atomic clock. We might even imagine that it keeps time perfectly. But this perfect clock is of no use to us unless there is some way for us to "see" the time it is keeping.

We know that the time of day can be quite conveniently displayed in our everyday world by hands that point to some particular time on the face of a clock. So let's suppose the hands on our super-clock are almost infinitely thin, so that we can read the time with almost perfect precision. Perhaps the hand we are observing is so small that we have to look at it under a microscope.

What is really happening when we read time in this way? Well, light is being reflected from the face of the clock back into our eyes, so that we can see the position of the hands with respect to the dial on the clock's face. That is, to read the clock we have to shine light on its face.

But there's a catch here. Light appears to flow continuously, like water down a riverbed. But water is not really continuous, and neither is light. Water consists of billions of molecules that are so small that they just *seem* continuous; and light really consists of little packets of light energy called *photons,* which bombard the clock face like rain falling on a metal roof. The photons cause the hands to jiggle ever so slightly—although our eyes are not sensitive enough to detect it, just as they cannot see molecules of water. In a dim light, the jiggling is small, and with a brighter light, it increases.

We are now reaching the heart of our dilemma. In a dim light the hands barely jiggle, but we can't see them very well. In a bright light we can see the hands better, but they jiggle more. There seems to be no way out; we cannot determine the location of the hands with absolute precision. Nature seems to have conspired to keep us from measuring the time with unlimited exactness.

Werner Heisenberg, a famous German physicist, was the first person to state in a precise, mathematical way, how measurements of time and other quantities such as energy and position are always limited. His observation was so important that it received his name, and is called the "Heisenberg Uncertainty Principle." Put rather simply, one form of the principle says that the more we know about *when* something happened, the less we can tell about *what* happened; and the more we know about what happened, the less precisely we can know when it happened. It is one of the most far-reaching and important discoveries ever made about nature, and there seems to be no way around it.

We introduced the Uncertainty Principle by showing how the position of the clock's hands are disturbed by shining a light on them. Although what we have said is true, this is not the whole of the Quantum Theory story. We have assumed that the hands have some definite position, but that we disturb their position when we try to see them better by turning a light on them. But do the ideas of position and time have the same meaning in the microworld of atoms and electrons that they do in our macroworld of rocks, trees, and houses?

If we think of atoms as being tiny, unbreakable billiard balls, then it may make sense to talk about an atom as being at a certain place. But suppose atoms are not like billiard balls. Suppose they are more like tennis balls, with fuzzy surfaces. Now it is more difficult to decide just where the tennis ball ends and space begins. Many experiments now confirm that atoms are not like billiard balls, and that they do not have sharply defined edges or surfaces.

Suppose we mentally roll an atom down a sloping board, and ask the question, At what time did the atom first reach a line drawn across the board near the bottom? We can answer this question only if we know precisely where the atom is; and we cannot know this because the atom is "fuzzy." In other words, it makes sense to ask about the position of an object that has a sharp surface or boundary, but not about one that has not. If we insist on talking about the position of objects that don't have definite edges, we've slipped back to trying to describe bats in terms of mice and birds. It makes *some* sense in terms of what we already know. But in the long run it creates more problems than it solves.

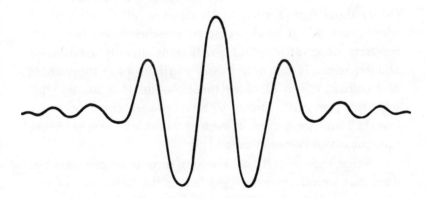

Wave representation of atom

Scientists have found that the only way to resolve some of the puzzles and contradictions in formerly accepted ideas is to describe the world of atoms and electrons in a new way —a way in which exact position and time are no longer completely meaningful. They now think of the atom as being something like a miniature wave with a crest that tails off on either side. This "wave" structure means that we can no longer think of the atom as a billiard ball-like object with a definite position. Rather we say that the most likely position of the atom is near the crest of the wave, but it could also be with less likelihood somewhere in one of the tails.

In a similar way, we cannot know the exact positions of the almost infinitely thin hands of our super-clock. The hands are made of atoms, which we now see as wave-like. So the hands do not have sharp edges and cannot be pointing to a definite location on the dial—which, of course, also consists of "wavy" atoms.

Like other theories we've been exploring, this wave-like nature of particles and matter seems contrary to common sense. But we must accept it because a "billiard ball" model for matter does not work. Thousands of experiments have been carried out that confirm the "waviness" of nature. And in the end, this is the best reason for accepting it—it works.

In the world of scientific study and experiment, any theory about nature must finally stand or fall on the basis of whether or not it works—that is, whether it explains and supports observations and predictions already established about nature. If the new theory explains away the puzzles and contradictions of older theories—and if it fits in with what was previously known to be true—then the new theory "works," and we accept it, even if "working" means giving up some of our commonsense ideas.

Space, time, gravity, and motion seem to be connected in ways that were not understood before the beginning of this century—or if they were understood, only dimly so. Scientists —and even persons who have little scientific background—in future years will probably look back on our present-day concepts of these relationships and see *our* understanding as "dim."

And so we return to our very first question that we posed in Chapter 1: What is time? We still don't know, although we've put it to some amazing uses and keep studying its mysteries and puzzles. Perhaps like the scientists before us, we'll just have to give up searching for an exact definition, and agree that time is simply what keeps everything from happening at once.

Glossary

ANTENNA—A metallic apparatus for sending and receiving radio, television, or other electronic waves.

BIT—A unit of information.

CLOCK—Any instrument for measuring or indicating time.

DOPPLER EFFECT—An apparent change in the frequency of sound, light, or radio waves that occurs when the source and the receiver are moving toward or away from each other. The frequency increases as they approach each other, and decreases as they move apart. Named for physicist Christian Johann Doppler.

FREQUENCY—The number of complete cycles or wave forms occurring in one second.

FREQUENCY STANDARD—A device that generates, with high accuracy, a signal at a particular frequency.

HERTZ (Hz) —A unit of frequency equal to one cycle per second, named for physicist Heinrich Hertz.

KILOHERTZ (kHz) —One thousand hertz or cycles per second.

MEGAHERTZ (MHz) —One million hertz or cycles per second.

MACRO—Refers to the very large.

MICRO—Refers to the very small.

MICROSECOND (μ-s) —One millionth of a second.

MILLISECOND (ms) —One thousandth of a second.

NANOSECOND (ns) —One billionth of a second.

NOISE—A random or persistent disturbance that interferes with the clarity of a signal.

OSCILLATOR—An instrument that produces a steady rhythm of swings or vibrations.

PARTICLE—A constituent of matter. Electrons, positrons, photons, neutrinos, antineutrinos and many other "small chunks of matter" to which scientists have given names are particles.

RADIATION—The creation and emission of waves and particles.

RESONATOR—The frequency-determining device in an oscillator or frequency standard.

SIGNAL—The sound, image, or message sent or received by telegraph, telephone, radio, television, or radar.

TRANSMITTER—Electronic equipment that generates and amplifies a carrier wave, modulates it with a message signal such as speech or tone, and radiates the resulting signal from an antenna.

Index

National Bureau of Standards—10, 19, 23, 45, 50, 51, 53, 54, 56, 57, 62, 63, 95, 96, 103, 110, 123, 125, 128, 131, 136

British Tourist Authority—15

J. B. Lippincott Company, *What Time Is It?* by I. Y. Marshak, 1932—17

Bulova Watch Company, Inc.—26

Gary Rehn—46

U.S. Department of Commerce—77

King Features Syndicate, copyright © Rube Goldberg—106–107

Bell Laboratories—142